FARMING SYSTEMS DEVELOPMENT IN FLOOD PRONE BASINS

S. Saran
M.M. Sinha
Ajay Kumar

2018
Daya Publishing House®
A Division of
Astral International Pvt. Ltd.
New Delhi - 110 002

First Published, 1998
Reprinted, 2018

ISBN: 978-81-7035-201-3 (Hardbound)

Published by : **Daya Publishing House**®
A Division of
Astral International Pvt. Ltd.
– ISO 9001:2008 Certified Company –
4736/23, Ansari Road, Darya Ganj
New Delhi-110 002
Ph. 011-43549197, 23278134
E-mail: info@astralint.com
Website: www.astralint.com

Acknowledgements

Successful organization of a major project requires time, effort and funds and support from professionals, individuals and institutions. We extend our hearty thanks to :

1. The Indian Council of Agricultural Research for their support to Farming System Research programme in India.
2. The Ford Foundation for the logistic, financial and professional support from the beginning to end of the project.
3. Rajendra Agricultural University, Bihar, Pusa (Samastipur) for logistic and professional facilities.
4. The resource persons Dr. A. Bottrall, Dr. B. P. Ghildyal, Dr. K. C. John, Dr. Nancy Auxin, Dr. Ruth G. Alsop of the Ford Foundation.
5. Dr. Stephen Biggs and Dr. D. Gibbon the FSR experts of University of East Anglia Norwich (U. K.).
6. The FSR team members Dr. R. C. Chaudhary, Dr. Kishore Kumar, Dr. N. Chaudhary, Dr. S. M. Ghufran, and Dr. S. H. Ahmad of Rajendra Agricultural University, Bihar, Mithapur Campus, Patna.
7. Mr. Ahmad Salman and Dr. P. Mishra of L. N. Mishra Institute of Social Science and Economic Change, Patna.
8. Vice Chancellors of Rajendra Agricultural University, Bihar Dr. K. K. Jha, Dr. G. Trivedi, Dr. P. N. Jha, Dr. J. C. Bakshi and Dr. K. S. Chauhan. Directors Research of Rajendra Agricultural University, Bihar Dr. P. P. Jha, Dr. P. Kumar and Dr. A. K. Srivastava.
9. All the experts from different faculties of Rajendra Agricultural University, Bihar who visited the project area from time to time and provided useful advice.

10. The experts Mr. B. Mandal and Mr. C. S. Das of R. K. Mission, Narendrapur, Calcutta.
11. Mr. S. K. Srivastava, Sr. Geologist, Geological Survey of India, Patna.
12. Dr. K. V. Raman, former Director, National Academy of Agril. Research Management, Hyderabad (NAARM).
13. A host of others who assisted us directly or indirectly in the planning and implementation of the programmes.

Foreword

High rainfall areas of Eastern India hold key to further marked increases in foodgrain production in the country. However, one of the most serious problems to be tackled in these areas is to contain the adverse effects of water logging. I am happy to note that Dr. S. Saran and his colleagues from the Rajendra Agricultural University, Bihar have attempted to bring out a book entitled "Farming Systems Development in Flood Prone Basins".

Dr. S. Saran has worked for a decade in a Farming System Research Project in Vaishali district of Bihar and his efforts have enabled him to come out with the present publication for sustaining the productivity of flood prone rainfed low lying systems (*Chaur*) of the Northern Bihar region. The efforts made by Dr. Saran and his colleagues are commendable and I congratulate them for this endeavour.

I do hope that this publication will be useful for planners, extension workers and the students.

(R. S. Paroda)
Secretary and Director General
Government of India
Indian Council of Agricultural Research

Preface

In complex, diverse and risk-prone rainfed agro-eco environment the conventional research, confined to experiment station, has failed to contribute anything significant for the disadvantaged farmers. Productivity of these systems is low and static.

In 1985 the Ford Foundation supported a Farming System Research (FSR) programme in Vaishali district of Bihar. The primary objective of the research programme was to develop viable and sustainable technologies for rainfed rice-based cropping systems for resource poor farmers. This objective was useful as starting point. Soon it became clear that the programme has to be re-designed so as to make it more relevant to the existing adverse agro-eco-systems.

The Farming System Research attempts to initiate the process of identifying and developing improved agricultural technologies with active participation of clients-resource poor farmers. The basic tenet of FSR is that research process must start by acknowledging the socio-economic and physical ecological context, governing farmers decision making. It seeks to link farmers indigenous knowledge system with formal service and technology to evolve new solutions.

The FSR programme functioned from 1985 to 1995. The FSR team of scientists was based at Agril. Research Institute, Patna. In 1987 Social Scientists of L. N. Mishra Institute of Economic Development and Social Change, Patna also joined the programme and helped the team in providing social science inputs. Consultants of the Ford Foundation paid regular visits to the FSR site and advised the FSR team in the development of the programme. One research scholar of University of East Anglia, Norvich, U. K. spent more than one year with the FSR group, Patna to study various aspects of the socio-economic problems of the area.

The FSR programme has demonstrated that (i) Participatory On-farm research in difficult environments is a feasible research management approach (ii) Technological options can be developed for varied agro-eco-systems through intensification of cropping and crop diversification and (iii) The technologies developed offer modest but significant pay-off to farm house-holds.

Success in generating alternative technologies has been achieved by a willingness to build on farmers knowledge, experimentation and participation in research, extension and evaluation.

We hope that our FSR experience and the results obtained by working in adverse agro-eco-systems in Bihar would be of great interest and use to the concerned persons and institutions working in On-farm research programmes.

S. Saran
M. M. Sinha
Ajay Kumar

CONTENTS

Chapter–1

Introduction

The Sen committee appointed by the Government of India in its 1984 report had drawn attention of the nation towards stagnation and low productivity in agriculture in eastern India and had stressed the need for concerted efforts to improve productivity in the region. The committee had elaborately discussed the institutional, economic and social constraints to effective technology development in eastern India. At the same time there was a growing realization that the top-down, transfer of technology approach (TOT) which had proved highly successful in promoting the Green Revolution in areas with assured irrigation in north-western India had failed to achieve any thing significant in the risk prone areas of eastern India. The lack of success of TOT approaches in rainfed areas led the scientists and administrators to hypothesise that in this kind of context the agricultural research process must begin with a clear understanding of farmers problems and practices which because of ecology and resource base, are highly variable and complex. In turn the one-way process of lab to land need to be replaced by a two-way iterative research process. This process starts with research involving farmers and their resources; feed back ideas for new research at on-station and on-farm experiments with the aim to producing multiple options from which the farmers could select the most promising for their own specific situations.

The Ford Foundation provided substantial financial support, for the on-farm research in rainfed lowland rice in seventees to certain agricultural universities/states in eastern India (including Bihar). The results from this programme were not according to expectation. The scientists followed the TOT approach. In early eightees the consultants of the Ford Foundation and agricultural scientists of eastern India decided that new approaches to research in rainfed eco-system should be attempted with key element 'Participatory research'. Thus Farming System Research (FSR) approach was initiated in Bihar.

The FSR strategy emphasises that the research agenda should be determined by well defined farmers need through an understanding of the existing farming system rather than its perception by the researchers. It is a holistic approach with the main focus on strengthening the partnership between the researchers, extensionists and farmers through extensive farm activities. It provides farming community an opportunity to influence the research agenda to generate technologies. The researchers in the FSR programme place importance to identification of constraints in existing production systems by adopting Farming Systems perspective.

During the early phase of the FSR programme initiated in 1985 in Bihar, the focus was on the specific objective of developing improved technologies for deepwater rice eco-system. Social scientists from university of East Anglia (U.K.) and L. N. Mishra Institute (LNMI) of Economic Development and Social Change played important role in redetermining the research thrust areas. Subsequently, farmers innovative ideas and suggestions led the researchers to explore modification in land-use pattern during the dry season or risk free environment *i.e.* period between post-water-receding and onset of monsoon and flooding. Soon it was realised that researches on crop diversification and alternative options to raise family income are essential. Research on animal sciences, fishery, agro-forestry, social science and allied subjects were taken up.

As the work progressed importance of institutionalizing the FSR approach was realised. It was realised that without active involvement of administrators and policy makers it was not possible to implement the development programmes in the area effectively. Policy dialogue exercises were initiated. Results from these exercises were highly encouraging. The policy dialogue exercises where farmers, administrators NGO's and FSR practi tioners came together to develop the FSR programme is unique in the sense that such attempt was never made any where in India, before.

In the project area located at Vaishali district of Bihar there are a number of inter-connected flood basins locally called *chaur.* A *chaur* is a low/deep water land created by frequent shifts in the rivers course originating in Nepal and characterised by poor drainage and prolonged water logging. The FSR site has complex agro-ecological and climatic conditions. The entire area is dependent on rainfall for agriculture production. The agro-ecosystems of the area are impacted by a variety of socio-economic factors including land distribution, occupation, household income etc.

Land and water the major resources of the area are being exploited fully by the local community. The resource utilization practices have been developed during the course of centuries and reflect the farmers wisdom and innovative capacity. While considering the rational behind the current status of resource utilization one has to keep in mind the (i) adverse agro-eco-system (ii) socio-economic factors (iii) share-croppers choice and (iv) risk-adjustment strategies.

The experience and insights gained by working with farmers as partners in technology generation programme presents a method of balancing the component and systems technology generation process. The method has been found effective in developing an integrated farming system in flood-prone regions. The regular monitoring of farmers risk-coping mechanisms and elucidating the underlying principles and role of risk allocation enabled the FSR team to adopt a gradual and incremental approach to evolving a portfolio of enterprises. Strategies to diversify enterprises which lower systematic risk and allow-risk-free seasons and sectors to be exploited have also been highlighted.

It is recommended that in a complex, dynamic and risk-prone eco-system, the FSR scientists should investigate ecological-based land use system of the farming community rather than being circumscribed by entry-point strategy. Adopting an integrated land and water resource management perspective has opened up new and interesting lines of enquiry. It also encourages a search for linkages among crop-livestock-fisheries-horticultural enterprises in a flood-prone region. Implications for methodological innovations are explored.

The FSR programme has demonstrated that (i) Participatory on-farm research in the risky environments is a feasible and appropriate research management approach. (ii) Technological options can be developed for limited and dispersed locations through intensification of cropping and crop diversification. (iii) The technologies developed and tested offer modest but significant pay-off to farm households.

The programme has made a modest contribution to changing the climate of opinion regarding the introduction of the FSR concept in India. The links to NAARM and the involvment of ICAR scientists in the earlier reviews in particular seem to have been important in this regards.

Our approach has been to focus on the following key elements in this book.

(i) A holistic view of the farmers circumstances—agro-climatic and socio-economic.

(ii) Classification of different client groups—resources and access to supplies and markets, specific technological problems.

(iii) Multidisciplinary *diagrams* and ranking of client's problems as the basis of the research agenda.

(iv) Participatory on-farm research for developing and testing technology.

(v) *Demonstrations* and monitoring of technology in the field.

(vi) Systematic participatory procedures for monitoring, reviews, evaluations, implication analysis etc.

(vii) Increased flows of information and effective methods to facilitate feed-back and liverage throughout the technology development system especially in terms of strengthening research and extension linkages.

Chapter – 2

The Project Area

2.1 Location

The Project area having its location at Berai, Talgarha and Belkunda villages of Vaishali district in Bihar, India is about 40 Km north east of Patna across the river Ganges. The work was started in 1985 for verification and modification of the Farming System Research models and for generation of economically viable technologies. The site is located at latitude 24°45' N, longitude 85°18' E and 57.8 m altitude (Fig. 1). The area has enormous social, economical and environmental complexities and comes in the category of complex diverse risk-prone agriculture.

2.2 Physical Characteristics

The area is characterised by hydrologically adverse basin lands locally known as *Chaurs*. There are about 400 *chaurs* spread over an area of 0.65 m ha in North Bihar (CWRS 1990; Thakur *et al* 1990). These basins (*chaurs*) are rainfed low deepwater lands, created by frequent shifts in the course of rivers, originating from Himalaya in Nepal. Soils range from sandy loam in upland to clay in lowland having neutral to alkaline pH values. Soils are rich in potash and nitrogen but poor in phosphate. The soils are highly fertile. *Chaurs* are of different shape and size where water accumulates for 4-6 months (Saran 1977; 1987). Flooding time, duration and depth are highly variable and is linked with time and quantum of precipitation during monsoon (June-October). *Chaurs* are characterised by poor drainage system. They are inter-connected and flood water in the *chaurs* of the project site, comes from as far back as 40 Km^2 north and passes on down stream through a series of interlinked *chaurs* (Fig. 2) to the south into Ganges (Saran *et al* 1992). The project area

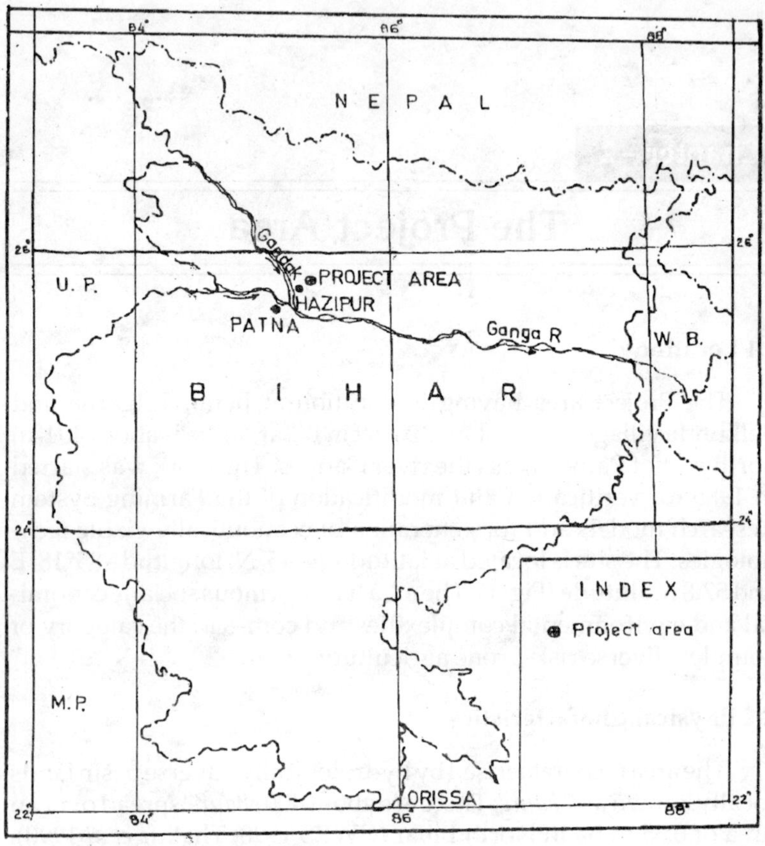

Fig. 1 : Map of Bihar showing location of the project area

falls in the mainstream of a natural drainage course between Ghagra (a tributory of river Gandak) and Ganges. The interlinked hydrology of a series of *chaurs* gives rise to a complex, diverse and high-risk (flood prone) ecology to the project site (John 1990). It has saucer-shaped depression with variable water regimes 5.0-400.0 cm (Fig. 3) with poor drainage system having different micro-ecological niches (Saran *et al* 1992 and 1993). Sudden flooding during July-August is annual feature of the area. Flood water recede pattern is

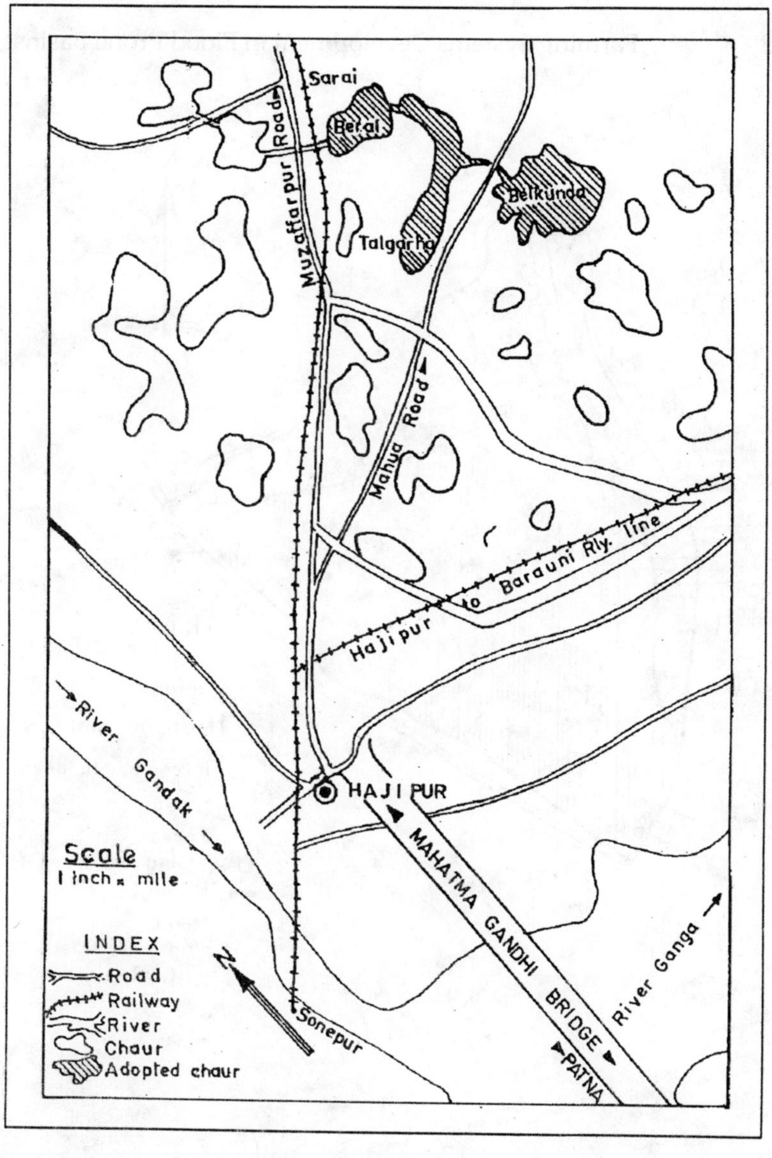

Fig. 2 : Inter connected chaurs and their location

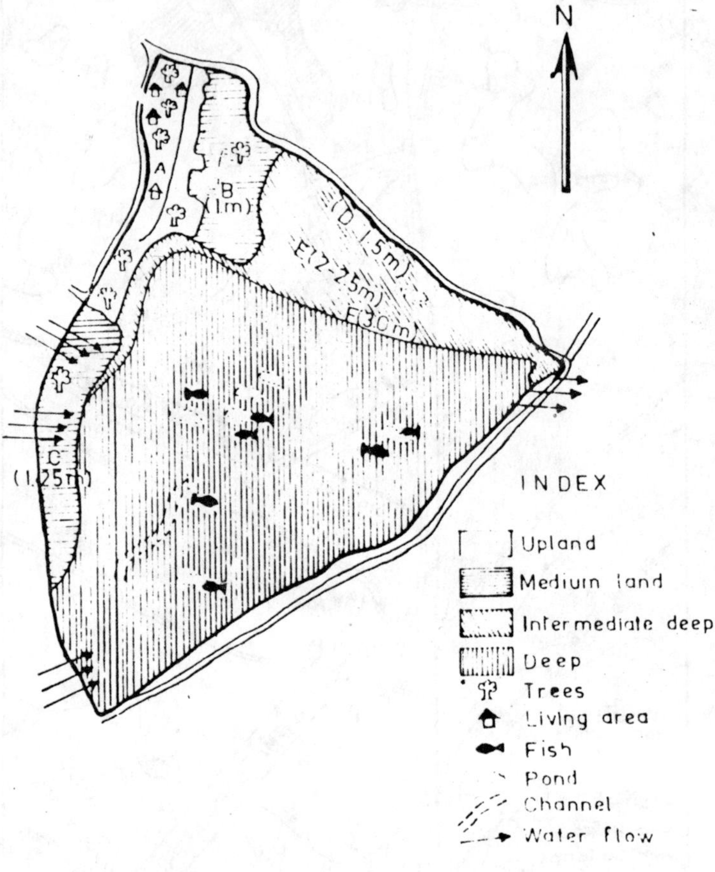

Fig. 3 : Micro-environments in the FSR site, Talgarha chaur

very slow. The water receding takes place during November to June, exposing the fields for cultivation. However, the receding pattern is dependent upon the water conditions.

Plate 1 : Flooded *Chaur* Area

2.3 Climatic and Hydrological Conditions

The temperature is highest in April-May, reaching upto 42°C and it goes down as low as 5°C in December-January. Average annual precipitation in the area is around 1200 m.m. The rainfall is highly uncertain and so is the flooding pattern. Month-wise water depth in the various land types in Talgarha *chaur* was recorded for four years during 1987 to 1990 (Fig. 6a-6d) which indicates variation in water regimes in different months and years. The erratic variable and uncertain hydrological conditions in the *chaur* have direct influence on agriculture and fish production.

2.4 Agro-ecological Features

There are two seasons for agricultural crops - monsoon (wet) season from June-July to October and winter (dry) season from October to March-April (Saran *et al* 1992). The monsoon crop season is more risk-prone than the dry crop season. The monsoon crops suffer invariably on account of flooding of varying depth and duration while in dry season, crops experience drought of varying intensities. Recycling of water from deeper niche to the upper portion of the *chaur* is also practiced traditionally to ensure harvest of winter (rabi) season crops. However, land topography hinders and limits the recycling of water for irrigating the winter crops. The hydrological situations-flood and/or drought give little scope for crop diversification. Rice is the major crop for monsoon, whereas, pulse (mungbean) is the predominant crop grown as mix-crop (Mungbean + Sorgham + rice). Water receding pattern from the fields is one of the important determinants for cropping pattern. In case of late recede (after March), raising of mungbean is not possible and there is no option but to grow deep water rice (DWR). The agro-environment of the *chaurs* are influenced by several factors like-precipitation, drought, water receding pattern, wind velocity, temperature *etc*. Flooding, drought and water receding pattern are very important agro-environments which influence the cropping system of the area, either alone or in conjunction with the hydrological factors and determine the crop-production. There is inter and intra-years variability in agro-environments of the area.

Plate 2 : Mixed crop of rice, Mungbean and Sorgham Spp.

Plate 3 : Bumper crop of local deep water rice "Darmi Jagar" during favourable season

2.5 Land Classification and Resource use Pattern

The land of the project area has been classified by the farmers, primarily on the basis of hydrology (water regimes) and topography of the fields. Farmers have broadly categorised the land into six types which are locally known as *Bheeth, Ucha chauri, Neecha chauri, Sarkurwa, Leev* and *Machgarha. Bheeth* land situation represents the upland, *chauri* (*Ucha* and *neecha*), the medium/intermediate lowland and *Sarkurwa, Leev* and *Machgarha* come in the ambit of deep water situations (Table 1). The *Sarkurwa, Leev* and *Machgarha* are grouped together and locally known as chaur. The farmers have identified ideal niches based on the performance of different land races of rice. The cultivar *Lakshman* land type is shallow *Baster* type is favourable lowland and *Darmi Jagar* is deep water ecology (Kumar *et al* 1990). The correspondence between land classification is presented in Table 1. Land situations of the Talgarha village has been depicted in Table 2 where the resource utilisation in the various land situations has been shown. How different ecological niches are utilised with the help of available resources has been presented in the transect of Talgarha chaur in Table 3. Integrated farming system for different micro-environment has been shown in Table 4.

Table 1 : Indigenous environmental classification of flood prone basins.

	Land types		*Rice land races*	*Water depth (m)*	*Ecological niches (as shown in Fig. 1)*
1.	Bheeth			0	A
2.	Chauri				
		Uncha Chauri	lakshman, rango	Upto 0.5	B
		Neecha Chauri			
			Bastar	Up to 1.00	C
			Bastar and Darmi Jagar	Upto 1.5	D
			Bastar and Darmi Jagar	Up to 2.0	E
3.	Chaur				
		Sarkurba	Darmi-Jagar	Upto 2.5	F
		Leebh	Darmi-jagar	Upto 3	G
		Machagarah	Darmi-jagar	Upto 4	H

a. Upland (*Bheeth*)

The fringe of the saucer shaped chaur is the upland and there is no accumulation of water here. Upland occupy almost 10 per cent of the total area (Fig 4). Upland (*Bheeth*) are used for constructing house, rearing cattle, growing crops such as tobacco, vegetables, maize pulses, oilseeds and for growing trees. The soils are light in texture and are flood free where crop management, ensuring good

Table 2 : Land Situations of the Talgarha Village

	Upland (Bheeth)	***Medium land (Uncha chauri)***	***Intermediary deep (Neecha chauri)***	***Deep (Chaur)***	***Drainage (Channel)***	***Pond***
Soils	Sandy loam	Loam	Clay loam	HC	HC	HC
Water resources	Hand pump well, tube well, pond	Tube well, lift irrigation from neecha chaur and chaur.	Pond, lift irrigation from chaur	Pond	—	—
Water accumulation sources	—	Rain water from catchment area, erratic rainfall, flooding from Ganges River.	Poor drainage and rainwater or rainwater from catchment area, flooding from Ganges River.	Poor drainage & rainwater or rainwater from catchment area, flooding from Ganges River.	—	—
Water regime	—	≥ 50 cm.	≥ 100 cm.	≥ 400 cm	≥425cm	≥ 500 cm
Crops	Tobacco, maize, moong, till caster, wheat, mustard sweet potato, onion, chili, tomato, ladies finger, pumpkin, brinjal, garlic, karela, cauliflower, mooli.	Rice, sugarcane, sesbania, moong, jowar, summer rice, maize, til, wheat, ladies finger, berseem,	Rice, jute, moong, til Jowar, maize, sesbania.	Rice, moong, jowar, sesbania, jute, til.		

Contd....

	Upland (Bheeth)	*Medium land (Uncha chauri)*	*Intermediary deep (Neecha chauri)*	*Deep (Chaur)*	*Drainage (Chaur)*	*Pond*
Trees	Mango, banana, jack-fruit, neem, peepal, palm, bamboo, babool, khajoor.	Bamboo, babool, palm, khajoor.	Palm, Khajoor	—	—	—
Livestock & poultry Fish *etc.*	Bull, cow, goat, poultry birds.	Captured fishing, snail and crab gathering	Captured fishing, snail and crab gathering.	Catching fish, snails & crabs Catching migratory birds	—	—
Other enterprises	Cultured fishing in ponds. Mat & net weaving, tobacco processing, pan making.	—			—	—

Latitude - 24° 45' N, Longitude-85° 18' E, Altitude-57.8 m Average rainfall-1200 mm/year, HC - Heavy clay.

Table 3 : Transect of Talgarha chaur.

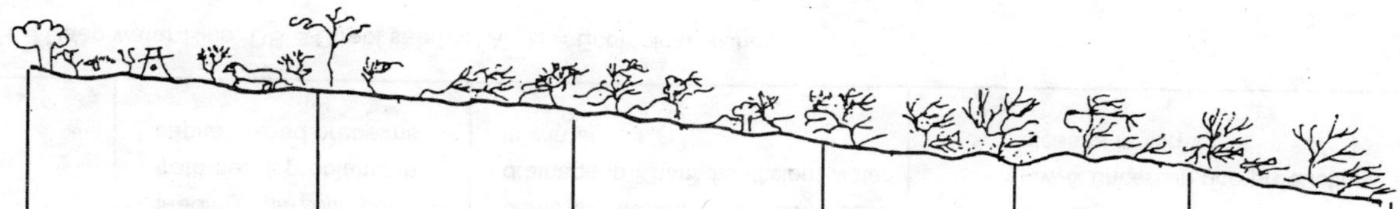

Land	*A* *Beeth*	*B* *Uncha Chauri*	*CDE* *Neecha Chauri*	*F* *Sarkurba (chaur)*	*G* *Leebh (chaur)*	*H* *Machargara*
Water regime		0.5	C 1.00 D 1.50 E 2.00.	2.5	3.00 Pond 5.50	4.00 Pond 5.00
Onset of submergence	No	Mid-August	Mid-August	Mid-July	Mid-July	Mid-July
Receding start		End November	Mid-January (C) Mid-January (D) End February (E)	Early March	End March	End April Continued stagnation
water resources	Hand pump, well, pond and boring.	Boring	Rainfall, rainwater from catchment area, flooding from river Ghaghara.			
Soils	Saline Soil	Clay		Heavy clay		Heavy clay
Crops	Tobacco, vegetable, oilseed, maize, mung, wheat, til, jute, rice, jowar, sugarcane.	Rice (trans), DWR + till + jute, suga r-seed	+ Mung + jowar + cane, wheat, oilseed	DWR (trans) or DS, jowar	DWR transplanted without puddling or direct seedling	

Contd......

	A *Beeth*	B *Uncha Chauri*	CDE *Neecha Chauri*	F *Sarkurba (chaur)*	G *Leebh (chaur)*	H *Machargara*
Trees	Mango, banan, jackfruit, neem, peepal, jamun, palm, bamboo, babul & khajur	Bamboo, palm khajoor, khar				
Livestock and birds	Bullock, buffalo, cow, goat, duck.				Migratory birds	
Fish, snail, crab	Rehu, katla, mirgal	Captured fishing, snail and crab				
Other enterprises	Mat weaving net weaving ving, tobacco processing					
Problems	Salinity, low productivity, diseases in animals, disea-se in fish, problems in goat feeding, marketing of ros-trata seed, problems in capital, weed probelms.	Rice brown spot, mealy bugs in rice, weed problems, rice damage by crabs, stem borers in rice, rat damage (kharif and rabi) poor drainage in Kharif insufficient water in winter		Mealy bugs and stem borers in rice, weed problems. Sudden submergence and poor drainage low & uncertain rice production disease in fish		

DWR = Deep water Rice, DS = Direct seeded, A - H = Ecological Niches.

Table 4 : Integrated Farming Systems for Micro-Environment of Chaur

Niches	*Kharif*	*Rabi*	*Summer*
A.	Maize/Rice	– Fingeriling rearing	Maize
		– Tobacco & vegetable or spices	Moong
		– Maize & Rajma	Vegetable
		– Oil Seed*	Sunflower*
		– Marigold*	
B.	Rice (Transplant)	– Wheat	– Rice (Transplant)
		– Bakla	– Maize + DWR
	Sugarcane		– Sugarcane + Moong
C.	Rice (Transplant)	– Wheat – Barley	– Maize + DWR – Rice (Transplant)
D.	Rice (Transplant)		– Rice (Transplant) – Maize + DWR
E.	Rice (Transplant) DWR + Jute DWR + Rostrata Moong + Rostrata Jowar + Rostrata	**	– Rice (Direct Seeding) – Maize + DWR
F.	Rice (Transplant) DWR + Jute DWR + Rostrata Jowar + Rostrata	**	– Rice (Direct Seeding) – Maize + DWR
G.	DWR + Jute Moong + Rostrata DWR + Rostrata Jowar + Rostrata	**	
H.	DWR + Jute DWR + Rostrata		

* Integrated with bee-keeping

** Integrated with Aquaulture

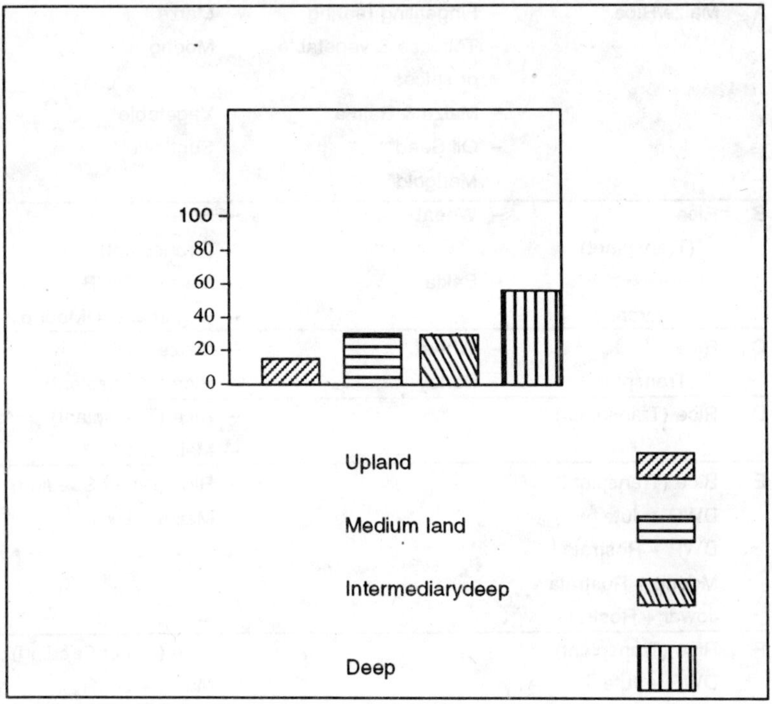

Fig. 4. : Distribution of land types in Talgarha village.

production is possible. Almost 70 per cent of the production comes from the uplands.

b. Medium/intermediate Lands (*Chauri*)

The chauri lands are dependent on monsoon rains and the flooding patterns are highly variable. Partial to complete flooding is dependent upon the low and heavy rains and there is inter and intra years variability in the flooding patterns of such lands. Water depth, ranges from 0-50 cm in 'ucha chauri' and 50-100 cm in the '*neecha chauri*'. Nearly 20 per cent of the area comes under the category of chauri lands having adverse crop environment and low productivity. Rice is the most important crop during the wet season (kharif) and is grown as mix-crop with Mungbean and Sorgham. During the winter (Rabi) season where the water recedes early wheat, mustard and pulse are cultivated in these lands (John 1988).

c. Deepwater Lands (*Chaurs*)

In the deepwater contiguous areas from intermediate land, the water depth ranges from 100-400 cm. This is the deepest part of the land mass and spread over about 15 per cent of the area. Water is an important resource. Almost 85 to 90 per cent of the *chaur* land remains water logged for 6 to 10 months. The cropping pattern in this area is largely influenced by the precipitation and receding pattern. Usually deepwater rice is grown during wet season (kharif). But in case of low rainfall and early water recede from the fields, mix cropping (DWR + Mungbean + Sorghum as fodder) is very common practice. In the deepest position of the *chaur* locally known as *'Machgarha land'* capture fishing is practiced by the fisherman which is the main source of their livelihood. However, fishing is practiced in other water logged areas of the *chaur* also. Resource-poor farmers also harvest crabs and snails from the lowlying areas for domestic purpose. There exists a great potential to raise substantially the socio-economic status of the resource-poor farmers by improving the fish culture (John 1989). Some of the farmers recycle the water of lowlying areas to irrigate their crops during dry season (Rabi). There is good scope to utilise the water resource which is available in abundance.

2.6 Water and its Resource use Pattern

During the year of normal rainfall the cropping pattern followed in Talgarha *chaur* has been presented in Fig. 5. This pattern holds good for other local chaurs also. Figures 6a to 6d depict the variation in the water regimes in the Talgarha *chaur* from 1987 to 1990.

Water is another important resource in the area and is available in plenty. About 85-90 per cent of the total area remains flooded for 6-10 months of the year. Such a large reservoir of water is very crucial for the farmers and influences their day to day life in a big way.

Traditionally farmers utilise the water reservoir for fish culture. A particular group of people are engaged in this profession. They (fishermen) are locally called *'Mallah'*. They are in this profession since centuaries and have acquired great knowledge and skill in the profession.

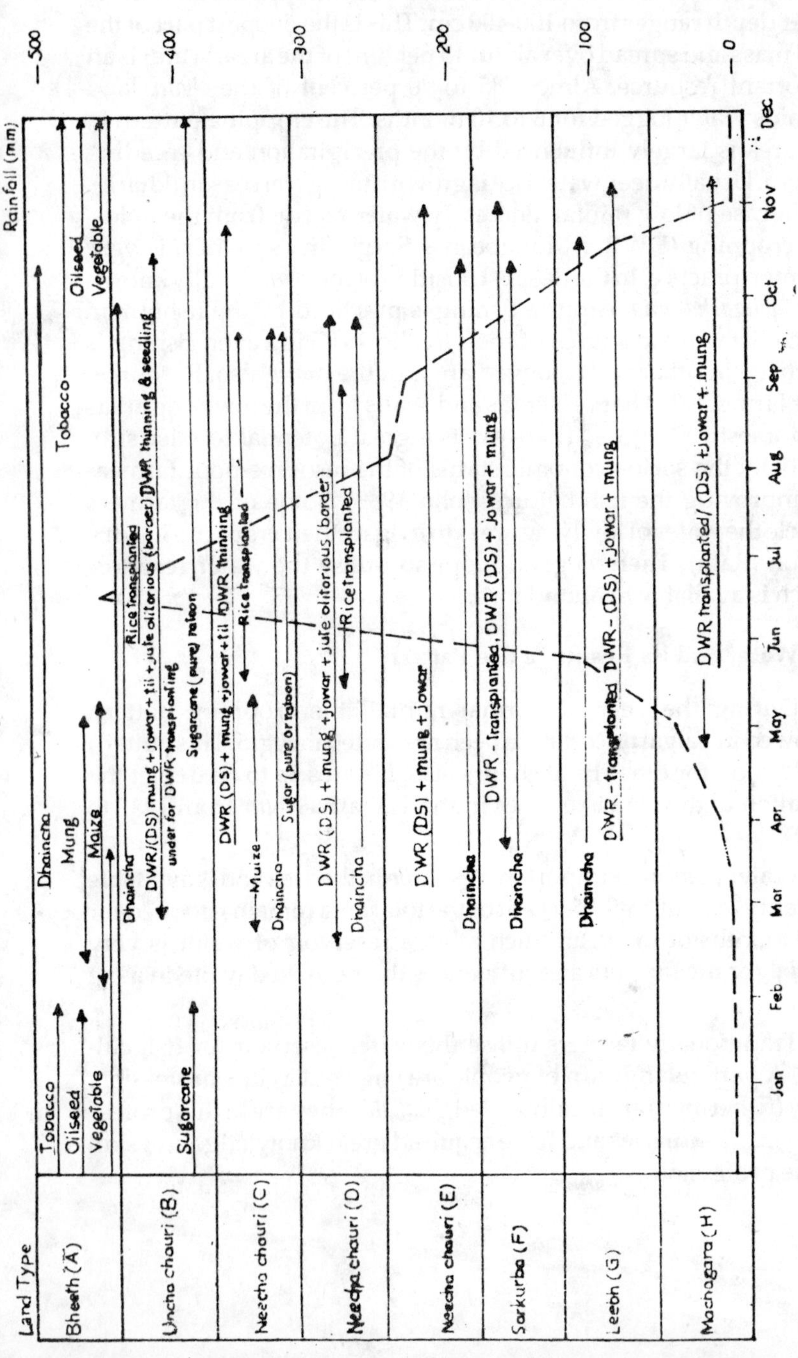

Fig. 5. : Normal rainfall and cropping pattern in Talgarha *chaur* DS-drect seeded, DWR-deepwater rice, A-H- ecological riches

Fishermen possess a wide range of equipments and tools for fishing during different period of the year. A good monsoon is boon for the fishermen but is bad for rice farmer. Fishermen practice capture fishing for their livelihood. Fish harvesting is done throughout the year except towards the last phase of monsoon (in September) period when flooding is at its peak.

Several varieties of fish are found in the *chaur*. Their harvesting time and period is also different. In October and November *Pothia, Kotri, Nanki, Jhunga, Kawai* and *Bhulla* are harvested. During December the harvesting of above mentioned varieties is continued in addition to *Dherba* and *Garai*. From January onwards big fishes like *Rohu, Katla, Buari,* and *Nami* are harvested. In the months of May and June water in the chaur comes down to the lowest level when *Singhee* and *Gainchi* are netted in the shallow water (all the fish varieties mentioned are in the local names).

The productivity of capture fishing is uncertain and low e.g. nearly 50 Kg/ha annually. Some of farmers dig up ponds in the chaur. During dry season (January to June) water remains in the ponds but it dries up or recedes in the rest of the *chaur* and *chauri*. This happens when monsoon is weak. In such a situation fishing remains confined to ponds mainly whose owners are mostly resource rich farmers. Certain fish varieties like *Rohu, Katla* attain good weight (upto 2 Kg). Growth of *Katla* variety is better. No feed is provided by the fishermen in the *chaur*. Marketing of fish is usually done by the women of Mallah community.

Snails and crabs occur in plenty in the *chaur*. Women and children particularly of the *Mallah* community augment their income by harvesting and marketing snails and crabs which also cause damage to the standing deep water rice crop.

Traditional method of fish harvesting is practiced by the fishermen. Different types of tools and equipments are used for catching various categories of fish. Cages are also employed to trap fish. There are a variety of cages locally known as *'Arsee'*. There is a range of *Arsees* which are called by different names *e.g. Ghana, Akardhi, Dwardhi, Tekua, Goal, Saraila* and *Argauli.* The fishermen also use different types of nets *i.e. Saurdhi, Dwanti, Tiar, Ghatgal, Chenua, Jalpar* and *Ghonijal* (used for catching the migratory birds). There are other methods of fish harvesting with the help of *Sahath,*

Bansi, Boats, Tapee and *Kurail*. The fishermen barricade the inlet and outlet of water in the *chaur* with the help of bamboo sticks or saccharum sticks and coconut coir known as Chilhaun.

2.7 Resource-use Decision Making Process

Three areas of participation in the decision making were considered (i) who initiated (ii) who influenced and (iii) who executed decisions.

Results of three important areas *i.e.* disposal of family resources, cash and kind expenditure and agricultural technology are presented here. In most cases female participation increases as the household becomes more involved in agricultural wage labour indicating that intra household status of women is highest in low income families *i.e.* they have greater control and authority within the family unit.

Within the context of decision making relating to agricultural technology women hardly ever carried out the decisions. However they increasingly initiated and were involved in discussions about agricultural technology as the households were poorer. Investigations into the disposal of family resources examined decision making in expenditure on major items such as agricultural equipment and animals. Again a general inverse trend between female participation in discussion and resource wealth of households can be seen. Female of rich households often have cash or gifts from parents over which they have some control. Influence in this area would affect expenditure on items such as smaller agricultural inputs including seed (variety and amount) fertilizer *etc.* It was realised that where share cash is at, the tightest level there would be the greatest competition for use of this resource and that male/female discussions could play an important role in the possible adoption rates of technologies.

These findings illustrate the fact that both men and women are involved in decision making with the ratio of women to men participation increasing as household income decreases. Higher percentage for female can be correlated to their direct participation in the activities that are the subject of the particular discussion. In the case of richer households this is high only in tasks carried out in the domestic sphere. As household income decreases women move further into participation of those activities that have a relationship to incoming goods (cash or kind). It is therefore apparent that there is a relationship between women, an intra household status and their participation in economic activities outside the domestic sphere. If the existing pattern of withdrawl of women to the domestic environment as household financial status grows and is continuous then increasing yields and overall household income is likely to result in loss of intra household status for women.

2.8 Rationale for Resource-use

Land and water, the two major resources of the area have been exploited fully by the local community. The resource utilization practices have been developed during the course of centuries and reflect the farmers wisdom and innovative capacity. While considering the rationale behind the current status of resource utilisation one has to keep in mind the following factors that influence the resource utilisation to a great extent.

(i) Adverse agro-eco-systems : Almost the entire agriculture is dependent on monsoon. Flooding and drought are regular features of the area. Hardly 10-15 per cent of the land is safe for agriculture. Rest of the area is subjected to varying degree of flooding every year. Period, intensity and time of the flooding is unpredictable (Fig. 6a to 6d). Similar is the condition when drought factor is considered. Under such circumstances agriculture is highly risky.

(ii) Socio-economic issues : Land distribution, high percentage of illitracy, poor household income, large percentage of land ownership with the high caste people (who are in minority) caste and class factors, majority of the households under the category of marginal landless or smaller farmers are the important socio-economic constraints that have influenced greatly in the development of the present status of resource use.

(iii) Share croppers choice : Almost 100 per cent share croppers are poor and come under the category of small farmers, marginal farmers or landless labourers. Due to their resource poor condition and risky and adverse agricultural environment they have no choice but to follow the traditional agricultural and fishing practices which are ideally suited for them and provide the maximum return under the existing agro-eco and socio-economic sys-

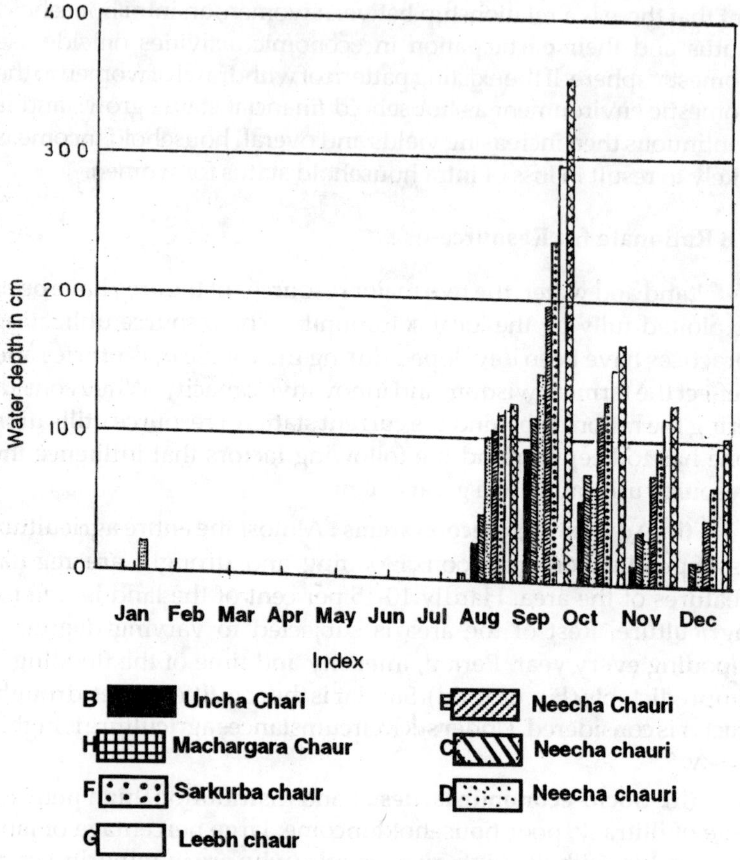

Fig. 6a : Month wise water regime in Talgarha *chaur* in 1987

tems. Modern technology has yet to be developed for such rainfed agriculture system. In fishery also the condition is similar but poor resource and social factors like common fishing in the *chaur* inhibit the development of fishery. There is great potential for fishery if the resources and modern technology reach them.

Participatory research expanded the spectrum of scopes in the development of horticulture, agriculture, agro-forestery, fishery *etc.* based on the resources available at village level.

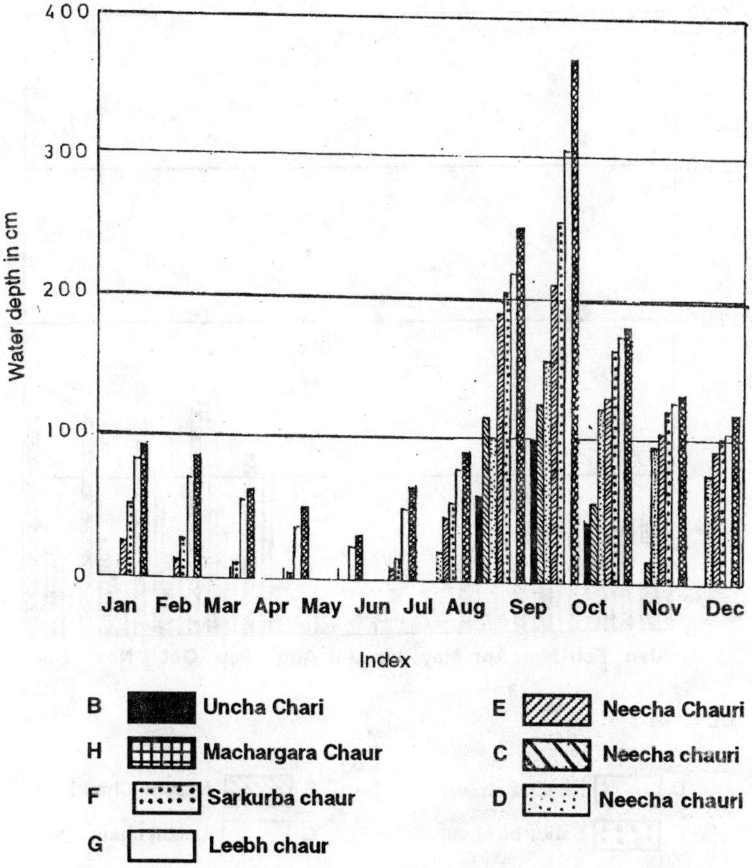

Fig. 6b : Month wise water regime in Talgarha chaur in 1988.

(iv) Risk-adjustment strategies : Various strategies (followed in the local systems) are developed in such a way that they fully take care of the risk involved due to adverse agro-climatic or socio-economic factors. Important features of such strategies are as follows :

Varieties/seed adopted by the farmers are highly suitable under local condition. In rice which is the most important crop, *Baster* (in medium lands) and *Darmi-Jagar* (in deepwater land) are the pre-dominant varieties. Both the varieties possess drought and

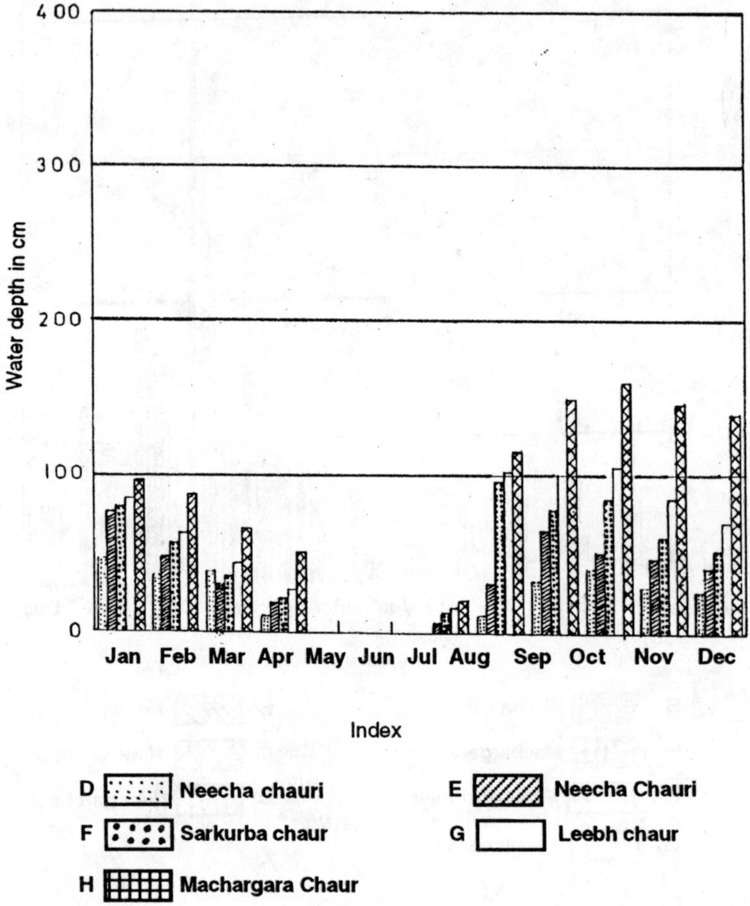

Fig. 6 c : Month-wise water regime in Talgarha *chaur* in 1989

flood tolerance and suitable duration and are resistant to a broad spectrum of insect pests and diseases. They possess elongation capacity so that they escape submergence. Their suitability under the risky local conditions can be judged by the fact that even after 10 years of on-farm research on rice based cropping systems, scientists could not develop or find a suitable variety as substitute

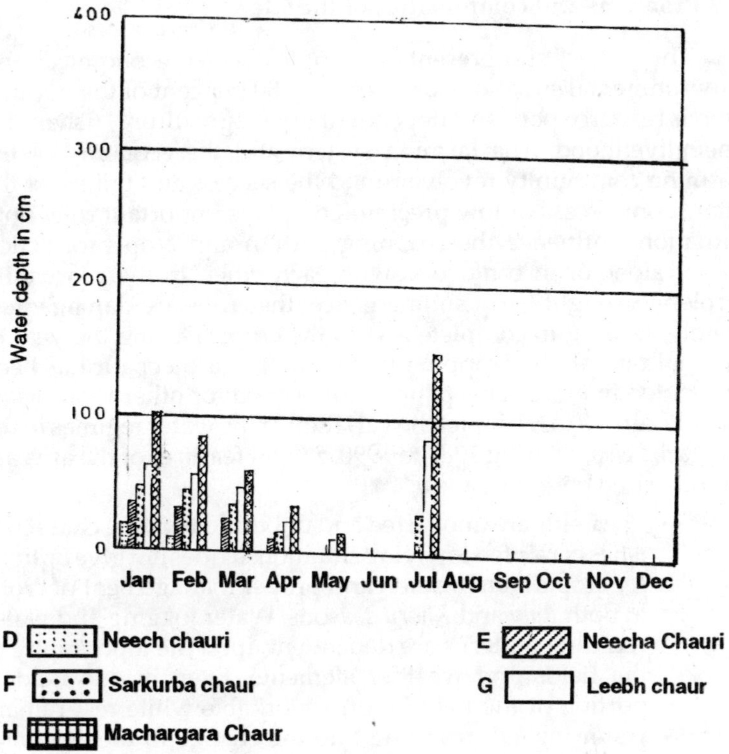

Fig. 6 d : Month-wise water regime in Talgarha *chaur* in 1990

for the local varieties. The same holds good for other crops *e.g.* mungbean, sorghum, tobacco, maize, pulses, oilseeds *etc*.

Agronomical practices including seeding time, seed rate, tilling method, harvesting time are so nicely developed that risks are minimised considerably.

Number of local fish varieties is large. This ensures fishing throughout the year. Fish varieties have different growing period. Thus for each month a particular fish variety is ready for harvest. Specific fish varieties have been identified for shallow water, intermediate water and deepwater conditions. This helps the fishermen to catch fish in various aquatic micro-environments either in drought or in flood.

2.9 Problems and complexities of the site

The project site presents a complex social, economical and environmental situations where nearly 80 per cent of the population is resource poor and dependent upon agriculture/fishery for their livelihood. The farming system and the economy of the farming community revolve round the success and failure of the monsoon. Excess or low precipitation plays important role. Both situations influence the cropping pattern and crop production either alone or in conjuction with each other. In the years with prolong drought/crop submergence, the crops are damaged severely, leading to complete loss of the crops. During the year of normal rainfall the cropping pattern in Talgarha chaur has been presented in Fig. 5. This pattern holds good for other local *chaurs* also. Figure 6a to 6d depict the variation in the water regimes in the Talgarha *chaur* during 1987 to 1990. Salient features of the area are summarised here.

— The fields are undulated and unbunded having characteristics of rainfed lowlying situation. It does not give options for crop diversification and proper management of crops in both *Rabi* and *Kharif* seasons. Water logging and water receding patterns are dependent upon the topography of the fields and weather elements. Even in one field, a portion of the field become workable while rest remain water logged, resulting into the change in the cropping system. The farmers face lot of problems in soil and crop management. Poor drainage system of the area is primarily responsible for the water-logging in *chaurs*.

— Drought and/or submergence are the major constraints to crop-production. Excessive drought and flooding situations make crop-husbandry less remunerative. In bad years the farmers face the shortage of food for themselves as well as feed for their cattles. Such situation imbalances the farming system and lead to the migration of resource - poor farmers for livelihood in the urban area.

— Considerable area remains water logged for most of the period of a year where capture fishing is practiced by the fishermen. In lowlying area no barricading to deleneate the area for fishing can be done because the *chaur* area become a lake during the monsoon period. The capture fishing in such situation comes under the common property right which can not be brought under individual property right for fishing. Moreover, there is the clash of interests between the farmers and the fishermen. The fishermen allow the inflow of water for better fish harvest, whereas the farmers do not allow the inflow of water to save their crops. This creates lot of social problems.

— Prolong water logging around the *Bheeth* makes the homstead area conducive to human and animal diseases.

— Sudden inundation of the chaur during *kharif* (July-August) causes irreparable loss to rice crop. On account of the hydrological variability and complexities, once in five years a good rice crop is harvested.

— In lean and delayed monsoon years and in absence of pre-monsoon rains, direct sown rice crop and other *rabi* crops suffer badly due to drought of varying intensities.

— The productivity of the area is highly variable and depends upon the time and distribution of rainfall during monsoon and also on the water inflow from the inter connected chaur system. In the year of low rainfall the farmers do not allow the water to pass down-wards by bunding the outlets for ensuring the water requirement of their crops. But in the year of excessive rains, area is completely submerged, leading to loss of crops. This situation brings mass querrel and litigation among the villages of the interconnected chaur system.

— Human induced interventions in the downstream during the past decade have impeded the natural drainage systems, thereby aggravating the problem of water logging. Moreover, the developmental process like construction of road, bunds *etc.* without taking into account the natural drainage of the agricultural land has increased the complexities of the basin lands (*chaur*).

— The set of technologies generated at On-stations are not effective and appropriate to meet the challenges of variable environments which are prevalent at the On-farms. The macro and micro environments of the risk-prone area can not be simulated at On-stations. Consequently, the rice varieties developed at On-station are not stable in performance in the area having complex environment. Aquatic weeds influence the crop yield. Most of them are location specific and detail study on their perpetuation in the system has not been made so far.

— Socio-economic status of the farming community plays important role to determine the growth in respect of agricultural and other non-crop production. The socio-economic structure has been discussed in chapter 3. Majority of the farmers are resource poor, landless, and share-croppers. They are dependent on the help of absentee farmers and they do not take up the long term production programmes.

SUMMARY

The project site is located at Berai, Talgarha and Belkunda villages in Vaishali district of Bihar, India. The area has enormous social, economic and environmental complexities and is characterised by inter-connected flood prone basins locally called *Chaurs*. The drainage system in these lands is extremely poor thereby causing water-logging for prolonged periods.

In this chapter physical, agro-ecological, geographical and climatic characteristics of the project area have been described along with the problems and constraints related to socio-economic and agro-ecological ddevelopment. Major resources available and their use patterns have also been discussed.

REFERENCES

Bottrall, A (1989). The Ford Foundation supported Farming System Research Programme in Eastern India. Paper presented for the 24 Annual Rice Workshop, Haryana. Agricultural University, Hissar, India.

CWRS, (Centre for water Resource Studies) 1990. Assessment of waste land problems in humid alluvial regions, a case study in North Bihar. Bihar College of Engineering, Patna University, India.

John, K.C. (1990). "An Overview of Farming System Research Programme in Eastern India". Paper presented at the ICAR-Ford Foundation Workshop on Farming System Research in Eastern India, India International Centre, New Delhi.

John, K.C. (1990). Faxiness evaluation of farming system research experiments in Talgarha Chaur.

Kumar, A; John, K.C.; Salman, A; Mishra, P. and Saran, S. (1990). "Policy Implications of Farming Systems Research Experience in Flood Prone Basins of Bihar". Best Paper Award 1990. Ist Asian Farming System Research and Extenstion Symposium, Bangkok, Thailand Nov., 10-12, 1990.

Kumar, A; K.C. John, and S. Saran (1990). "Sustainable Agril. Development in Chaur. Linkages among policies, transactions and ecological risks". Paper presented at the workshop on linking Farmer Participatory Research to Planning and Implementation 6-9 March, 1990. University of Kalayani, W. Bengal, India.

Saran, S. (1977). "Progress of deep water rice research in Bihar, India". Proceedings 1976 deep water workshop 8-10 November, Bangkok, Thailand.

Saran, S. (1987). "On-farm research in Rainfed Lowland Rice in Bihar". Paper presented in the international Deepwater Rice Workshop, Bangkok, Thailand.

Saran, S. (1990). "Farming System Research in Vaishali, Bihar, India". Paper presented at the ICAR-Ford Foundation Workshop on Farming System Research in Eastern India-India International Centre, New Delhi.

Saran, S; Mishra, P; Kumar, A; and Salman, A. (1992). "FSR in Risky Agroecosystems". Journal for Farming System Research Extension Vol. 3 No. 1 pp. 145-157.

Saran, S; John, K. C; Kumar, A; Salman, A; Sinha, M M. and Mishra, P. (1993). "Integrated Farming System Methodology for High Risk Ecological Zones". Vol. 1 No. 4 pp. 463-477.

Thakur, R; A.K. Singh, R. Singh and R.S. Singh (1990). "Survey of Deep water Rice in Bihar" (1987-1990). Rajendra Agricultural University, Pusa, India (Monograph).

Chapter – 3

Socio-Economic Profile

Traditionally, social scientists have played a 'critical' rather than a participatory role in agricultural research. In the past they have been brought into conduct adaption, constraint, impact and consequences studies. On other occasions they have been employed to carry out bench mark surveys so that they can evaluate the project at later date.

In FSR the role for social scientists is changed. They participate as a member of the research team to influence the direction and content of the programme. Involvement of social scientists in this new role is one of the major challenges (Biggs 1985).

Socio-economic constraints play important role in determining the generation and pace of the development of technology, its adaption and in turn raising the socio-economic status of the rural population.

In early stage of the project (during eightees) social scientists from the University of East Anglia, Norwich, (U.K.) were associated actively. In the late eightees consultants of The Ford Foundation were source of directions to pushing the FSR programme. Social scientists of L.N.M. Institute of Economic Development and Social Change, Patna, also joined FSR scientists team of RAU for conducting the socio-economic studies.

3.1 Study of East Anglia University Group on Systems Component

Dr. Ruth Alsop worked and collected data under a research programme from the area (Belkunda village) where agricultural scientists of FSR team were examining the possibilities of new rice technologies. Clients were identified as resource poor producers, operating in rainfed situations. Inter and Intra households information showed that time allocations, decision making and labour use directly assist efficiency within the confines of this research domain. It was also used to demonstrate how, by looking at other elements of the system (*e. g.* livestock and feed), the productivity of more than one system component can be enhanced. Relationship between these components illustrates the underlying principles of the programme with equitable and efficient results which would be integral part to an agricultural research effort. The analysis outlined first the social and economic organisation of the village in relation to agricultural/livestock production and resources. Secondly, it focused on the intra household differentiation across groups. Finally, information on agricultural research interests emerged from this social and economic exercises.

(i) The Village Disaggregated

There were eighty seven households in the village which were divided into nine caste groups, 5 scheduled, 3 backward and a Muslim community. Sample selection was based on agricultural labour use or connection with agriculture/livestock. Households not falling into these categories were ignored. A total of two hundred and fifty four people were included with a male : female ratio of 1: 0.8.

Five income strata were devised based on households net annual income. There was a positive correlation between high income group and size of holding (Table 5). The richest 9 per cent of cultivating households had 74 per cent of the land and the poorest 31 per cent possessed 4 per cent of the land. The lowest income group farmers were cultivating 68 per cent of land on a share crop basis meaning thereby that 50 per cent of gross product accrues to the landlord. In the low income group 16.6 per cent of their land was partially irrigated, no such farmer had fully irrigated upland. Almost 9.5 per cent of the total sharecropped land was of this type. Conversely in income group I, 23 per cent of land owned and cultivated was fully irrigated and used mainly for growing tomato and tobacco as cash crops, 11 per cent was partially irrigated (wheat, maize and vegetables) and 21 per cent was *chaur*. Of the land rented out by the highest income group, 53 per cent was *chaur*, 29 per cent fully irrigated upland and 18 per cent was rainfed shallow water situation which was used mainly for paddy in kharif season and wheat, gram and mung in rabi. Land ownership and access was therefore showed in terms of gross amount available and in terms of the land quality as measured by returns and security of that return.

Table 5 : Cultivators and Holdings over Income Strata.

Income Strata	*Average size of Holding (Ha)*	*Percentage cultivated land in Village*	*percentage cultivators in Village*
I	7.48	74	9
II	1.02	8	9
III	0.41	9	28
IV	0.26	5	23
V	0.15	4	31
		100	100

Original stratification of the sample was carried out along the lines of net use of agricultural labour. Four strata were formed (i) households who were net hirers of agricultural labour (ii) households who were net users of family agricultural labour (iii) households who were net hirers out of family agricultural labour and (iv) households who used no agricultural labour for cultivation but who owned land or livestock. This stratification proved to be a relatively good indicator for the resources wealth of households. Disaggregated in this way households displayed similar patterns in ownership and access to livestock as to land.

(ii) Income and Livestock Ownership

Positive correlation was observed between high income group and high incidence of ownership of livestock. In labour use strata 3, (net hirers out of family agricultural labour), suggested that although 90 per cent of households have access to livestock but 86 per cent of these animals were taken on a share basis. These households receive half of the gross income only. In case of full income of households there was an increase in the incidence of share animal (Table 6).

(iii) Sources of Income to Household

Tables 7a and 7b illustrate the importance of various sources of income to the households in different strata. For example, 62 per

Table 6 : Ownership of Access to Livestock by Different Resources Groups.

Labour Use Strata	*Total*	*Percentage of Strata with Livestock*	
		Owned	*On Share*
I	100	100	0
II	100	89	0
III	14	90	86
IV	86	50	14
Income Strata		**Percentage Animals on Share**	
I		0	
II		0	
III		29	
IV		21	
V		50	

Table 7a : Ranked Sources of Household Income

	Percentage of Village		
	Primary	*Secondary*	*Tertiary*
Agricultural/Livestock Production	17	38	0
Agricultural Wage Labour	31	5	1
Shop	12	0	0
Petty Trading	13	8	1
Tea Stall	2	0	0
Skilled Trade	15	0	0
Service	2	0	0
Profession	5	0	0
Government Employment	3	5	0
Other	0	1	3
TOTAL	100	57	5

Table 7b :

	Income Strata				
	I	II	III	IV	V
Percentage Income from agricultural production	72	44	12	23	25
Percentage Income from Wage	19	37	58	53	59
TOTAL	91	81	70	76	84

cent and 72 per cent of households in low income groups IV and V respectively were net hirers out of family labour and 53 per cent of income is derived from agricultural wage work in income group IV and 59 per cent in income group V. Alternatively 72 per cent of the income for group I was derived from agricultural/livestock production and 19 per cent from non-agricultural wage sources. Wage labour was an important source of income for a large number of households. Agricultural wage was paid both in cash and kind. Some time wage was linked with other factors such as renting of land, loans or security of future work. Computation of market wages value was inappropriate. Out of the earning through wages there was generally little surplus to invest in any form of asset. Ownership of capital goods was directly related to household annual income. There was little scope to poor families for investing because they had the only readily available resource is human capital. Rich families producing cash crops possessed full control of the generated income. Although there exists a strong market awarneness the mode of production is not purely capitalised. Labour was the only commodity with poorer families to sell at market value. Labour was initially identified by cultivators as problematic during peak agricultural times. Investigations were therefore carried out on time allocation, using a Random Instant methodology and hired labour use (monthly recall).

In respect of amount of time men and women of different economic groups were spending as agricultural wage labour and at what time of the year, it was noted that July/August and November/December were periods of maximum intensity for both sexes. Neither male nor female members of income group I were involved in agricultural wage labour. There were only two occurances of income group 2 males working for agricultural wages and no female of this group was involved. The remuneration improved households income but skewed the pattern of female wage employment.

Table 8 shows the pattern of labour hired over the busy six months viz; crop growing, harvesting and processing period. The discrepancy between the hired overtime and days was due to the movement of labour to other locations during time of high demand. The wages of these migrant workers was higher compared to the local agricultural labour. Hirers of labour reported that there was

Table 8 : Average Number Days of Agricultural Labour Used per Month by Sample Households in Labour Use Strata

Activities	Days per Month (Male = M Female = F)						
	June	*July*	*August*	*September*	*October*	*November.*	*December*
M	10.2		14.4	2.1	0.5		
F	0.5						
M	2.0	32.0					
F							
M	12.3	32.3	9.6	0.5			0.3
F							
M	31.2						
F							
M	8.8	6.4	0.6		0.4		
F							
M	1.0		0.4	0.5			0.4
F							
M	0.1	1.0					
F							
M		80.3	34.3				
F							
M			2.4				8.3
F							
M			0.3				
F							
M			0.3				
F							
M				6.5	0.1		1.0
F							
M						18.0	69.3
F							0.7
M							
F							8.3
M						25.6	33.3
F							
M							
F							
M							1.3
F					5.0	2.8	
TOTAL	66.1	152.0	62.3	9.6	5.5	46.4	122.9

Activities

1. Weeding
2. Sowing Rice
3. Ploughing
4. Tobacco
5. Land Maintenance
6. Applying Fertilizer : Chemical/Organic
7. Sowing Vegetable Seeds
8. Transplanting Rice
9. Sowing/Planting Tomatoes
10. Applying Weedicide
11. Irrigating
12. Preparing Seed Beds
13. Harvesting Rice
14. Beating Rice
15. Weeding Tobacco
16. Sowing Wheat
17. Transplanting Tobacco.

an increase in work if engaged on a contract basis. During peak period the labourer were moving out to work, suggesting that mobility had to some extent broken the bonds and was changing the nature of labour relations.

(iv) The Household Disaggregated

Each household in the sample was disaggregated by sex and age, and details on decision making, time allocation and labour use were collected accordingly.

(a) Labour and Time : The Time Allocation data not only gave an indication of the amount of time was spent on various activities at different times of the year, but also of who did what, An underlying assumption of this analysis was that there were four ways in which the sexual division of labour manifested. The tasks were *sex sequential, sex segregated, sex supportive* and *sex substitutable.* With this framework it became easier to see the dependency network within households labour relations and to ascertain the constraints or opportunities existed for changing or introducing new practices. Table 9 outlines seventeen major time use categories and the percentage of time spent by men and women between the hours of 5.00 and 22.00 over seven months. This information was categorised by income strata where different pattern of activities occurred. More detailed stratification of these major categories into 17 activities served to identify specific activities and showed specific relationships within a major area.

Women spent on an average 5.7 per cent of their time in the major category of livestock and poultry (Table 9). Within this category of time use, 4.1 per cent of their time was spent in collecting fodder. However disaggregation by income groups showed that, over the total time period, strata-I women spent no time in fodder collection while women in strata II spent 1.2 per cent, in strata-III 4.4 per cent, 4.1 per cent in strata-IV, and 11.3 per cent of their time in strata-V. In case of men pattern is slightly different having 1.1 per cent in strata-I, 2.0 per cent in strata-II, 1.7 per cent in strata-III, 1.9 per cent in strata-IV and 5.0 per cent in strata-V, the time spent for arranging fodder by men.

For the richest households, animal care was a *sexually segregated* work. In strata-II in July and August, women collected fodder when

Table 9 : Time Use By Men and Women Over Seven Months

Major Activity Category	Percentage Time spent Men	Women
1.	8.8	5.7
2.	0.7	0.0
3.	0.1	0.0
4.	1.9	0.2
5.	1.0	0.7
6.	1.4	0.0
7.	0.2	0.2
8.	1.4	1.4
9.	0.2	1.5
10.	15.0	2.3
11	1.6	0.1
12.	2.1	20.3
13.	10.6	7.8
14.	6.6	2.0
15.	4.0	6.6
16.	46.3	47.0
17.	7.1	4.3
TOTAL	100	100

Activity No.	*Activity Category*	*Activity No.*	*Activity Category*
1.	Livestock and Poultry	10.	Participation in Local Economy
2.	Land Maintenance and preparation	11.	Construction (Household).
3.	Fertiliser/Manure	12.	Domestic Activities
4.	Crop Growth and Maintenance	13.	Child Care
5.	Harvesting and Storage	14.	Education
6.	Vegetable/Fruit	15.	Social/Religious
7.	Hunting and Gathering	16.	Leisure
8.	Manufacturing	17.	Other
9.	Food Processing		

fodder was in scarce supply and males were spending a lot of time in raising and maintaining crops. This indicating that activities were *sex substitutable* or *sex integrated*. Only 0.5 per cent of the sample indicated that women worked as wage labourers, suggesting that fodder collection activity in income strata-V was sex substitutable. Males assumed heavy agricultural wage work leading the females in strata-V to spend more than average time in fodder collection as 7.0 per cent against an average of 4.3 per cent and 13.1 per cent against an average of 11.3 per cent.

The overall pattern was that the females were increasingly involved in fodder collection as household income falls. The substitution of labour between the sexes was governed by the decrease in income. This indicated that conflict between the social ideal to keep women in 'purdah' and economic pressure caused deviation at the time of stress. Activities associated with rice production were divided into three categories (i) Field located work (crop growth and maintenance, harvesting, drying, bundling and carrying), (ii) Household located work (threshing, winnowing, drying of grain, storing) and (iii) Processing work (cleaning, parboiling, post boiling drying, other activities). The following results were obtained relating to percentage of time spent during seven months.

Category I		*Category II*		*Category III*	
Male	Female	Male	Female	Male	Female
9.5	1.3	0.6	1.6	0.2	5.5

Female participation decreased as work location moved away from the home particularly in the women of high income group. Group-I females did not take field work while group-5 spend 0.6 per cent of their overall time in the fields. Male field activities increased as income decreased. Income group-I spending 0.4 per cent of their time while group-II 1.0 per cent, group-III 1.8 per cent, group-IV 1.0 per cent and group-V 4.4 per cent. This demonstrated the sex sequential nature of time allocation for rice production. The efficient execution of work was dependent upon different sexes sequential performing different roles. This disparity was observed in time spent by men and women in the execution of work because of the fact that some of the activities was performed by one sex only. Such tasks were sex segregated. Labour substitution occured in the poorer group where females took the sowing and transplanting of rice. The analysis of labour was used to illustrate the links between fodder collection and rice production. Shortage of labour during rice harvesting, processing and storage suggested that less time was available for collection of animal feed. Sex supportive roles like cooking, child care, religious/social obligations and other female domestic works were important. Major part of the time of women was spent on non-agricultural works. However, these works had direct relation having supplimental role over the farming functions.

The situation in the rice production was the overall concept of sex sequential activities, backed by a social ideal of sex sequential with evidence to support the assumption that sex supportive roles were important to the system. Cutting across this chain of interdependency, the economic forces influenced to varying degree the amount of task integration and substitution occuring between the sexes.

(b) Decision Making

Three areas of participation in decision making were considered (i) who initiated, (ii) who influenced and (iii) who executed. The results summerised in Table 10 were felt to be of immence importance to the agricultural system. In most cases female participation increased as the households were increasingly involved in agricultural wage labour (proxy for resource wealth of household). The intra household status of women was highest in low income families, *i.e.* they had greater control over their lives and greater authority within the family unit.

In the context of decision making relating to agricultural technology women hardly had any role to influence the decisions. However, they increasingly initiated and were involved in discussions about agricultural processes as the households were poorer. Inverse trend between female participation in discussion and the resource wealth of the household was the important feature. Cash and kind expenditures looked at more regular outgoings. This influenced expenditure on items such as smaller agricultural inputs including seed (type, variety and amount, fertilizer *etc*). However, it was felt that where cash availability was low there was high competition for use of this resource and male/female discussions were playing important role in the possible adoption rates of technologies.

Both men and women were involved in decision making with

Table 10 : Womens participation in Decision Making

	Percentage involvement by Women									
	Stage	*Initiation*			*Consultation*			*Execution*		
	Area	a	b	c	a	b	c	a	b	c
Labour	1	9	70	3	18	74	6	9	5	0
Use	2	0	76	6	20	78	9	0	22	3
Strata	3	25	84	6	50	89	23	50	58	0
	4	50	90	3	50	92	6	0	70	3

Area a Disposal of Family Resources
b Cash and Kind Expenditure
c Agricultural Technology

the increased ratio of women to men participation as household income decreased. Under such circumstances women moved further into participation in decision making that had direct relationship to incoming goods (cash or kind). There was relationship between intra household status of women participation in economic activities outside the domestic sphare.

3.2 Study of L. N. Mishra Institute on Socio-economic input

Social Scientists of L. N. M. Institute, Patna, India conducted socio-economic studies separately in the project area. Their study included agrarian and social status, demographic features, income, occupation, inter-household dynamics, availability of agricultural inputs and role of women in local Farming Systems.

A sample of 120 household was selected on the basis of multistage stratified sampling method. Farm activity of the selected households were regularly monitored and studied through interviewing and participant observation. Thirty two items of work were selected for the study which were grouped into five categories for analysis.

(i) Social and Agrarian Status

The agro-ecosystems of the area are impacted by a variety of socio-economic factors including land distribution, primary occupation, household income of different economic groups *etc.* A remarkable feature of the land holding in the project area is that 90 per cent of the total cultivable land is water logged (shallow and deep) and only 10 per cent is upland (Fig. 3 & 4). More than 90 per cent of the land holding is with the high caste households who constitute only 30 per cent of the total households (Salman *et al* 1988; Fig. 8). Caste and class factors which are important constituents of Indian society are highly pronounced in the area. Out of the total households, 80 per cent are either marginal, small farmers or landless labourers and nearly 20 per cent are resourceful (Table 11). Occupationally 65 per cent of the total households are cultivators, 15 per cent landless labourers, 6 per cent artisan and 14 per cent depend on service, business and trade in the area (Fig. 9).

Table 11: Distribution of sample households by resource condition and caste (in per cent)

Village/Resource Caste/Condition	*Resource Poor*		*Resource*	
	Land - less	*Marginal & small land- holdings*	*Advantaged*	*Total*
Belkunda				
High Caste	10.0	10.0	7.5	27.5
Low Caste	22.5	37.5	12.5	72.5
Total	32.5	47.5	20.0	100.0
Telgarha				
High Caste	5.0	20.0	5.0	30.0
Low Caste	12.5	52.5	5.0	70.0
Total	17.5	72.5	10.0	100.0
Berai				
High Caste	2.5	12.5	15.0	29.2
Low Caste	7.5	40.0	22.5	70.8
Total	10.0	52.5	37.5	100.0
Mean				
High Caste	5.8	14.2	9.2	29.2
Low Caste	14.2	43.3	13.3	70.8
Total	20.0	57.5	22.5	100.0

1. Marginal farmers have less than on hectare land and small farmrs have between 1 and 2 hectares.
2. Resource advantaged include large-scale, business people and others whose incomes exceed Rs. 6000/year.

(ii) Income, Occupation and Inter-household Dynamics

Farm income of the people included income from crop produce, milk (cows and buffaloes), fish harvesting, goat rearing *etc*. Non-farm income consisted of wages, salaries and profits from business and trade. However the major sources of income of the resource poor households consisted of agricultural and non-agricultural wages, share-cropping, tobacco processing, fishing, dairy products and goat rearing. The important feature of such a complex and high risk farming system is low percentage of sustainable yield.

Agriculture income from the deep water land is highly uncertain and depends on a variety of factors such as rainfall, floods and drought. Almost 40 per cent of the agricultural income comes from the 90 per cent of the land which remains water-logged during the major part of the year and 60 per cent of the income comes from upland which is only 10 per cent of the total land area (Saran *et al* 1992).

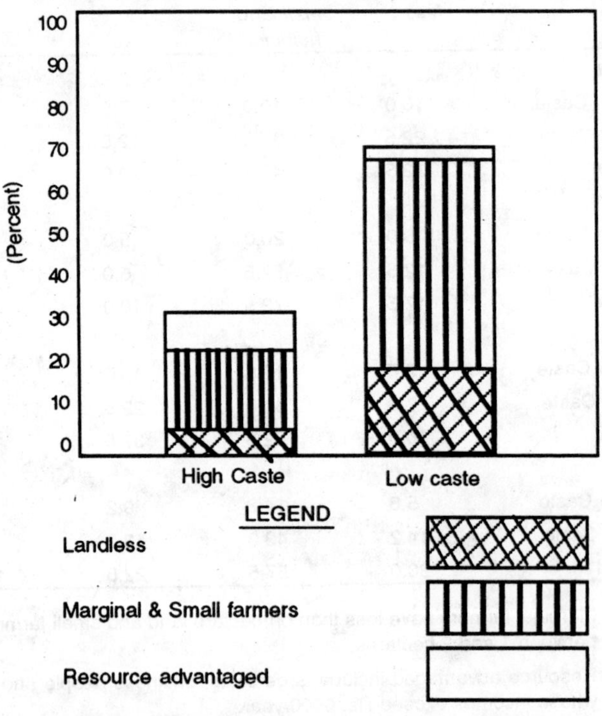

Fig. 8 : Distribution of households by caste & resource

(iii) Livelihood Analysis

Besides growing crops and vegetables, people of the area are engaged in animal husbandry, fishing (including crab and snail harvesting) and mat making. Cows and bullocks are reared by a small section of the population and there is scarcity of fodder crops (Salaman *et al* 1988). The problem of fodder is more acute during the water logged period. Most of the fodder requirement is met by

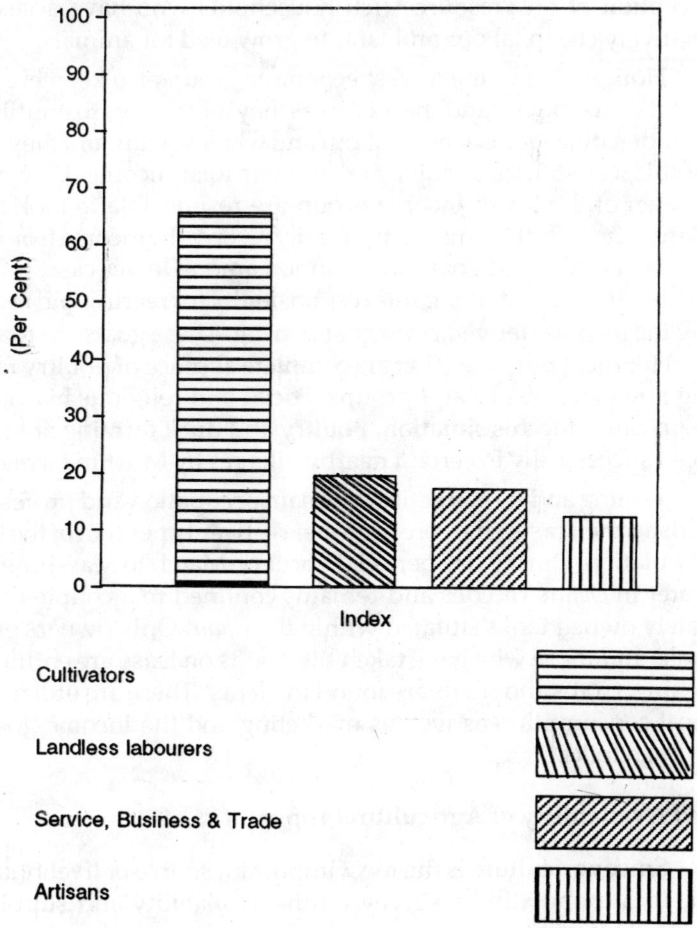

Fig. 9 : Occupational distribution of households in Talgarha village

cutting the grass from fallow land. Farmers have found the supplementary source of cattle feed in the form of banana biomass, bamboo leaves and wildgrass of the *chaurs* (locally called *karmee* and *mokna*). Growing of fodder crops is not practice, mainly due to limited availability of free land which is utilized mainly to grow food or cash crops for domestic consumption or marketing. Absence of fodder crop cultivation in the area is also linked to the uneconomic return from the local cattle breeds. Big animals are mostly reared by the lower castes of the population with the exception of few resource rich households who have access to relatively cheap labour and land to grow feed for animals.

Households in the lowest economic group also keep big animals both on owner and share basis as they have access to unutilized or underutilized household labour and whatever amount they earn by this, constitutes a major part of their total income. Low caste women of the lower income group are responsible to look after these animals but the men in the family receive the income from this source. Local breed goats are common among lower caste households with female having sole responsibility for rearing and receiving the income derived from goat rearing. These goats are mainly sold for meat purpose. There is complete absence of poultry farming among all castes and groups. Social and religious bias is the main cause for this situation. Poultry and duck farming are practiced successfully in certain nearby villages by Muslim farmers.

Fishing and its marketing are main occupation and profession of fishermen caste who constitute nearly twenty per cent of the local population. During dry period (February-March to May-June) the water in *chaurs* recedes and remains confined to a couple of privately owned tanks situated within the *chaur*. Only owners of the tanks and those who have taken the ponds on lease are entitled to fishing. Crabs and snails are found in plenty. These are utilized for local consumption as well as marketing and the income goes to women folks.

(iv) Availability of Agricultural Inputs

Since agriculture is the most important source of livelihood of the local population a survey on the availability and supply of agriculture inputs (which are very important for agricultural growth) was conducted. Only one resource that is the storage facility is 100 per cent available with the farmers. Farmers get more than half of their requirement of agricultural labour in time. They get hardly 10 to 20 per cent of their seed and fertilizer requirement in time. More than 80 per cent of the inputs are either not available in time or nor available at all. Availability of seed and scientific know-how are almost negligible. Seven resources *e.g.* irrigation, seed, fertilizer, plant protection, scientific know-how, marketing facility and agricultural labour are available in varying degree in time but with difficulty (Table 12).

Table 12 : Availability of agricultural inputs

Resource	*Easily available*	*Timely available with difficulty*	*Timely not available*	*Not available at all*
Irrigation	—	5	12	83
Seed	2	17	28	53
Fertilizer	—	22	26	52
Plant Protection	—	2	38	60
Scientific know-how	2	8	25	65
Agril. Implements	—	—	33	67
Storage facility	100	—	13	87
Marketing facility	—	2	60	38
Agril. Labourers	54	13	17	16

(v) Role of women in Farming System

The age-old feudal system still has deep influence on the society of the area. It is believed that women should not work in the farm and should confine themselves to their houses. Caste has great influence on the role of women in agriculture. The women of upper caste take very little part in the farm operations. However, they do take part primarily in post harvest operations which are done inside or near their houses. Low caste women play significant role in the agricultural system.

There are several castes in the region, which have been grouped into two *i.e.* upper caste and lower caste. Seventy per cent women of the area belong to the low caste and 30 per cent to the upper caste. Caste specific roles of women in agricultural activities are presented in Table 13. Upper caste women take part in only 30 per cent of the total farm activities, whereas, low caste women participate in

Table 13 : Women's Participation in Farm or Farm-Allied Activities

Activity	*Per cent of women Involved in this activity*	
	High caste	*Low caste*
Seed selection	—	83
Field preparation	—	10
Sowing/transplanting	—	58
Manuring	—	2
Weeding/inter-culturing	—	84
Plant protection operations	—	—
Irrigation	—	—
Harvesting	—	73
Thrashing	—	59
Grain cleaning	31	90
Storage	31	88
Grain marketing	—	71
Hulling/milling	—	25
Fodder harvesting	—	75
Animal feeding	12	75
Dairy management	9	75
Milking	—	—
Fish marketing	—	23
Fishing	—	11
Crab/snail harvesting	—	58
Crab snail marketing	—	23
Vegetable cultivation	—	65
Vegetable harvesting	97	98
Vegetable marketing	—	4
Mat-making	—	71
Mat-marketing	—	10
Sewing	46	8
Knitting	79	15
Maize shelling	100	100
Mung bean harvesting	27	86
Goat rearing	3	79
Tobacco cultivation	—	—

86 per cent of the farm activities. There are certain items in which high percentage of women of both categories participate *e.g.* in vegetable harvesting. In activities like sowing and knitting the upper caste women's participation is much higher. Grain cleaning, storage and mungbean harvesting are activities where upper caste women do participate, although at a lower percentage than the lower caste women.

Table 13 presents a holistic picture of the women's participation in different day to day activities with respect to caste . To go a step further, attempts were made to present class and caste specific distribution of women's participation in various activities. The findings are summarised in Table 14. All activities have been categorised into six classes. The percentage of women's participation in different activities presented in Table 13 reflects the number of activities in each sub-category. When we move from the lower economic class to the higher one, we find a general decline in women's participation in all activities irrespective of caste. We observed very high percentage of resource poor lower caste women's participation in all the farm activities. No high caste women's participation is observed in the pre-harvesting activities and activities related to fish, crab and snails harvesting and marketing.

Table 14 : Distribution of Women Participating in Various Activities According to Caste and Economic Class (in percentage)

Type of Activities	Resource Poor			Resource Advantaged		
	High	*Low*	*T*	*High*	*Low*	*T*
1. Pre-harvesting activities	Nil	68.3	41.2	Nil	27.3	9.4
2. Post-harvesting activities	40.7	92.7	72.1	19.0	45.5	31.3
3. Animal husbandry	25.9	85.4	67.7	19.0	36.4	25.0
4. Activities related to fish crab and snails	36.4	12.5	Nil	41.5	25.0	Nil
5. Vegetable growing and other related activities	81.5	97.5	91.2	52.3	54.5	53.1
6. Other activities	55.6	78.0	69.1	33.3	63.6	43.8

The high percentage of resources poor low caste women's participation in various agricultural operations is mainly due to their poor economic and social status. They have little choice left for their very survival and meeting day-to-day consumption requirements. On the other hand, upper caste women are economically and socially better off and they participate in the work according to their socio-economic status and needs. To increase agricultural output in the area, it is essential to initiate specific programme and policies for women to enable them to play their roles more effectively and thereby improve their own and their families economy. In most of the agricultural farming operations, from seed selection, sowing to harvesting, threshing, cattle rearing and dairy management, low caste women play a very significant role.

Women do not have any role, in plant protection operations, irrigation, milking and tobacco cultivation. There is a traditional belief that women's participation in tobacco cultivation would yield a poor tobacco crop. Harvesting mungbean, goat rearing, cattle maintenance, preparing cow dung cake (for fuel), harvesting snails and crabs are exclusive area of work of lower caste women.

3.3 Implications for Research and Development

The information on labour use, time allocation and decision making gave the dynamics of relationship within a farming system, both within and between households. It was the matter of importance to know the relationship between outcome of research and issues of equity and efficiency. An underlying assumption is that the second cannot exist without the first.

Equity concerns were divided into two crude categories corresponding to inter and intra household. The former, was related to the avoidance of economic polarization within a society and the latter focussed on the issue of status within the household which were linked by efficiency. In terms of efficiency within the village, it was inefficient, particularly in aggregate welfare terms. It was also inefficient to utilize capital inputs. Labour in this case had low alternative market options and was available and substitutable. Within the household the decrease in womens status as household income increased suggesting that they had less control over the welfare of the whole household. Secondary sources recorded similar cases where female status has declined as they lost control over economic decisions even though gross income increased.

Different groups within the village have different needs and will be affected by new technologies in different ways. This applied to both economic and sexual variables. Social and economic information can help to identify intended clients and can also point to relationship that may be affected by intervention between groups. In the agricultural research programme under discussion here the clients had already been identified as resource-poor cultivators. For the household at the lower end of the income scale agricultural wages played major role in household earnings. Therefore, if technological change is likely to affect the hiring of labour, analysis of richer households and particularly the dynamics of their resource allocation is required for agricultural research focusing on the poor. In the agricultural wage market there were also ramifications for changing labour patterns. Higher demands would mean juggling time allocations, which in turn could affect the adoption of recommendations. Bearing this in mind it seems that research into new rice varieties and techniques would be likely to have inopportune effects. There would also be a higher probability of adoption if attention was paid to the staggering of timing of peak activities.

Time was the major dimension were sexual variables played a role. Both men and women are part of the production system. Females direct involvement in the agricultural/livestock system was in fodder collection, rice processing and storage. From the information gathered it appears that women are likely to be withdrawn from wider economic participation of the domestic sphare as houehold earnings grow. Livestock appears to be a possible area of women involvement, giving some economic control. Even in income group-2 the pattern of sex task substitution was that females were collecting fodder when demand for male agricultural wage labour was high. In richer households milk was readily available. Concentrates were used as supplementary feed. Feed was the constraint to milk production. Even banana biomass was used as a source of cattle feed. This suggests that inter cropping of fodder with rice could be the useful direction for research investigation.

Meat was also a costly commodity with goats as the primary source. Goats were predominantly the concern of the female member of lower income group. Mortality rates in goat was recorded to be high and most of these animals were kept on a share basis obtained from richer households. Improving feed for these animals could be one way to augmenting and ensuring output and in turn earnings of the low income group.

Low caste/muslim groups were keeping chickens who fell in the low income strata. Priority on improvement in quantity of eggs and poultry meat, could be one of the approaches to increase the income of the poor section of the society.

Since income from the water logged land is highly uncertain and is influenced greatly by rainfall floods and drought, it is highly desirable to undertake on-farm research on crop diversification. Crops suitable for dry season from November to July should be identified. Trials on fodder crops, oilseeds, pulses, cereals, vegetables and sugarcane should be taken up on priority basis. There is plenty of scope for the development of animal husbandry. Improved cattle breeds of cow, buffalows and goats deserve greater attention.

Research on fish culture should be taken up on top priority basis. This will have a big impact on the socio-economic condition of the farmers.

There is plenty of water available in the area. Researches on water management would benefit the farmers immensely.

Highest priority should be given on the improvement of drainage system of the area. This would go a long way in increasing the agriculture production in the area.

Supply of agricultural inputs is bad in the area. This problem has to be taken up at the district level administrative authorities.

Certain specific research areas were identified *e.g.* to develop improved mungbean variety having synchronous flowering so that harvesting could be done only once instead of 3-4 times as practiced now. Development/introduction of improved tomato cultivation which is one of the major vegetable cash crop in the area needs attention.

Researches on few other areas like improved harvesting processing, storage, would not only reduce crop loss but would also benefit women who are more involved with these operations. Besides improved mat and basket making would benefit women.

SUMMARY

The study of Dr. Ruth G. Alsop reveals that components within the rice production system in the village (Belkunda) surveyed are related in such a way that intervention causing a shift or change in one will automatically cause a change in another. Two major components of the rice production system, rice and livestock have been examined using analysis of intra and inter household information to show that cause and effect relationships differ over sexual and economic groupings. Four types of labour relationship have been developed and used to explain and understand the patterns of time allocation and decision making.

Within this framework time allocation demonstrated the nature of labour relationship both within the household and local economy. Likewise the time allocation and hired labour use information has given a picture of demands made upon time for particular activity and household during certain parts of the year. Decision making questionnaires revealed patterns of control over resource allocations.

The use of such information has shown the way it can trace the relationship between different elements within a production system. Therefore, areas for intervention and valuable information suggest that manipulation of one specific aspect can be used in positive way to develop strategies to increase the productivity of another aspect.

Placing emphasis on disagregation over sexual and economic variables it is aparent that research activities can ensure efficiency at the investigative and implementation stages. Both direct beneficiaries and secondary interest-groups can be identified. The respective needs and available resources of each can be ascertained. An awareness of different implications for these groups therefore means that equity issues can also be considered as an integral to an agricultural research programme.

Socio-economic studies conducted by L.N. Mishra Institute revealed that the agro-eco-system of the area is impacted by a variety of socio-economic factors including land distribution primary occupation, household income of different economic groups *etc.* The study brought out interesting information on social and agrarian status income, occupation and inter-household dynamics, livelihood analysis, availability of agricultural inputs, role of women in farming system *etc*. Salient features of the findings are summarized here :

— Agro-ecosystem of the area is highly adverse to agriculture and is characterised by water-logging and extremely poor drainage system which is the most important problem of the area.

— Income of the resource poor farmers can be augmented considerably by diversification of agriculture.

— Availability of agricultural inputs in the area is very poor.

— Women play key role in agriculture but their economic condition is lower than men.

— In order to raise socio-economic level of the women, the activities as hereunder are desirable :

* Introduction of improved breed of goats.
* Improvement in fish, crab and snails farming.
* Training on manufacture of mat and basket.
* Improvement in the harvest and post harvest technologies.

Diagnostic exercise of the socio-economic and agro-ecological characteristics of the project area generated useful informations which helped the FSR practitioners (i) to develop better understanding of the varied conditions and needs of farm families (ii) facilitated prioritization of the research items and (iii) develop and modify their on-station and on-farm research agenda.

REFERENCES

Acharya, Meena and Lynn Bennett (1983). Women and the Subsistence Sector : Economic Participation and Household Decision Making in Nepal. World Bank staff working papers No. 526.

Bigg, S.D., (1985). Report on Rice Based Farming Systems Research for Resource Poor Farmers of Eastern India" School of Development Studies, University of East Anglia Norwich (U.K.).

Chabala, Charles, and Robert G. Nguiru (1985). Intra-Household Dynamics and FSR/E in Zambia. University of IIlitnois at Urbana Champaign, U S A.

Carloni, Alice (1983). Integrating Women in Agricultural Projects : Case Studies of Ten FAO-Assisted Field Projects, FAO, Rome (Swaziland study).

Carloni, Alice (1983). Technical Report on Socio-Economic Constraints to Land and Water Development for Inland Valley S amps in Sierra Leone, FAO/ UNDP (SIL/80/010) Rome.

Farmer, B.H. (1997) ed. The Green Revolution : Technology and Charge in Rice Growing Areas of Tamil Nadu and Sri Lanka. MacMillan London.

Feldstein, H. (1986). Intra-Household Dynamics and Farming Systems Research : Conceptual Framework. Draft. From discussion of the Advisory Committee to the Intra-Household Dynamics and Farming Systems case Studies Project.

Grosvenor-Alsop, Ruth (1986). Rice, Meat and Milk System Components in North-east India. Paper presented in the International Farming Systems Research Symposium, Kansas State University, U.S.A.

IRRI (1983). Women in Rice Farming : Appendix Recommendations of the Women in Rice Farming Systems Conference (pp. 525-531).

Kumar, A. John, K.C., Salman, A., Mishra, P. and Saran, S. (1990). "Policy Implications of Farming Systems Research Experiences in Flood Prone Basins of Bihar". Best Paper Awards 1990. Ist Asian Farming System Research and Extension Symposium Bangkok, Thailand, Nov., 10-12, 1990.

Mishra, P., Ahmad Salman and Singh, C. (1988). Differential Class-Specific-Women Participation in Farming Systems of Chaur Area in Bihar. Farming Systems Research Newsletter (EIFSR Network, Rama Krishna Mission, Vol. II, No. 2).

Norem, Rosalie Huisinga (1986). Intra-Household Dynamics and FSR/E : Conceptual Framework (Concepts and Definitions).

Overholt, Catherine, Mary B. Anderson, Kathleen Cloud and James E. Austin (1985). Gender Roles in Development Projects, Kumarian Press, Connecticut.

Pradhan, Bina and Lynn Bennett (1981). The Role of Women in Hill Farming Systems. Paper presented in Seminar on Appropriate Technology for Hill Farming Systems, Kathmandu, Nepal.

Salman, A., Kumar, A., Mishra, P., Saran, S. (1988). "Caste and Class specific Differential Role of Women in Farming Systems of Bihar, with Special reference to Vaishali Chaur ; Paper presented at the International Farming System Symposium, University of Arkansas.

Saran, S. and A. Kumar (1988). Role of Women in Rice-Based Agricultural System in Vaishali District of Bihar. Paper presented in FSR Workshop, Ranchi, Bihar.

Saran, S. (1987). On-farm Research in Rainfed Lowland Rice in Bihar. Paper presented in the International Deepwater Rice Workshop, Bangkok, Thailand.

Saran, S., Mishra, P., Kumar, A., and Salman, A. (1992). "FSR in Risky Agro-ecosystems" Journal for Farming Systems Research-Extension. Vol. 3, No. 1. pp, 145-157.

Saran, S., John, K.C., Kumar, A., Salman, A. Sinha, M. M. and Mishra, P. (1993). "Integrated Farming System Methodology for High Risk Ecological Zones". Vol. 1, No., pp. 463-477.

Chapter–4

Participatory Technology Development Process

4.1 From commodity based to resource based research strategy

Radically different farming systems are developed in conventional agricultural research, using state of the art technologies and integrating crop, livestock and trees species in a modal farm at the experiment station. Inter-dependence among enterprises is acknowledged and taken care of in such a farming system (Ming 1980; Rangaswamy 1989). On farm trials are viewed as means of validating and demonstrating such integrated farming systems (IFS). However, the highly structured nature of such on-farm trials; inhibits farmers from experimenting with IFS and adapting them to their bio-physical resources and socio-economic conditions, precluding potential modifications and adjustments in the system. Often such interventions are not sustainable even in optimal production environments. The point of departure for developing an IFS is the recognition of the complexities of Indo-Gangetic alluvial flood basins (*chaurs*) and the diversity of responses of the local communities. We contend that farmers role in balancing the components and systems technologies within their bio-physical and socio-economic context becomes more critical and increasingly cost effective as environmental complexities increase and farming systems become multifaceted. The dominant commodity strategy has to field to a resource based strategy in order to blend components judiciously with systems technology. The objective being to evolve a diversified portfolio of enterprises as protection against ecological risks.

In such circumstances, conventional approaches that try to

identify management treatments for maximising out-put over time and space get unwieldy and unrealistic, while acknowledging that the commodity research strategy and the resultant integration of enterprises through component technologies would continue its way over national agricultural research systems (NARS) of developing countries. We argue that on-farm research with a farming systems perspective provides opportunities to complement commodity-based strategies with resource based research strategies. Besides farmers participatory research encourages farmers to evaluate research priorities and suggest re-orientation. It is the farmers themselves who hold the key to developing, evaluating and validating these systems (Okali and Sumberg 1986; Atta-Krah and Francies 1987; Ashby and Pachio 1989).

4.2 Problem Ranking and Research Prioritisation

The key question in deciding the farmer participation in the technology generation process is the control over the process of problem definition, research prioratisation, design and implementation of experiments (John 1990). Three criteria for problem ranking were considered e.g. distribution of the problem, importance of an enterprize in the local farming system and seriousness of the problem. Ranking on these criteria were obtained with the help of an informal survey (PRA) of target group of farmers. Alternatively FSR team members and the farmers jointly debated and ranked the problems. These exercises gave insights about the criteria to be employed as well as precise definition of the problems. Ranked problems were then examined with the help of system diagram to understand farmer perception of bio-physical causes and socio-economic constraints and their interaction.

4.3 Selection of Collaborator Farmers

Initial thrust was on deep water ecosystem and addressed research needs of the rice based cropping systems of disadvantaged farmers from this designated area. Subsequently the prevailing heterogeneity in the village ecosystems was studied. The whole village concept enabled the researchers to exploit variabilities in ecological as well as socio-economic condition.

Locating experimental plots were done through ecological/ agronomic mapping based on local knowledge. Farmers were consulted at research prioritazation phase.They were also briefed about the concepts of on-farm trials, variables were considered for testing. Farmers probable evaluation criteria on trials were also consulted. Pre-experimental briefing sessions helped the scientists in classifying mutual expectations and the role of FSR process.

Typologies of On-farm research

4.4 Implementing experiments, monitoring and farmer evaluation

On-farm research with Farming System Research perspective is substantially different from conventional on-farm trials, though there is tendency among researchers to use these two terms interchangeably. Most on-farm-trials/demonstrations are conventionally viewed as means of validating a given package of technology. The FSR team proceeded with a different objective of on-farm research. That is to identify opportunities in the agro-ecological conditions given the current production system and to develop individual technical innovations which would match with the existing production system.

Several interesting typologies of on-farm research have been put forward by researchers. Hidelebrand and Poey (1985), Asby (1986), Biggs (1989), Water-Bayer (1989), Garrity (1990) have tried to synthesize four major streams of on-farm research based on the following attributes : who designs the experiment : who manages it : and who implements it. The variants are

(i) Researcher implemented experiments
(ii) Farmer–implemented experiments
(iii) Farmer participatory experiments and
(iv) Farmers indegenous experiments.

The alternatives in on-farm experiments are :

(i) Observing and making inferances from farmer initiated experimentation and
(ii) Observing and making inferences from researcher - planned experimentation.

During course of the participatory research programme FSR practitioners attempted and gained experience from all the above types of experiments but emphasis was laid on exploratory testing followed by refinement and validation. The design of such trials was simple compared to researcher implemented trials.

One of the important consideration in the on-farm research was to maximise the variability *e.g.* ecological, socio-economic and managerial. This had significant implications in organising on-farm trials. Farmers environmental classification system was utilized to identify contiguous plots.

Monitoring of the on-farm trials was done regularly by researchers to record the variables of the status of environment and crop population (vegetative stage, growth stage at flowering, yield component, pests and diseases etc.). Besides block monitoring was carried out (by researchers and farmers) from time to time. Block monitoring was done on plots having similar soil and micro-climate but cultivated by different farmers. This information helped in comparing existing variability in crop management. Block monitoring was also carried out in different ecological niches.

Evaluations of the standing crop at critical stages was carried out to know the farmers views. This was done for eliciting directly from farmers what they think of the proposed technological innovation independent of researchers assumptions. The essence of farmer participatory methods was to obtain direct point of view at each stage of research diagnostic, problem ranking and prioritisation, farm trials and monitoring.

4.5 Technology Development Process

On-farm research began with the deep water rice varietal trials only with specific objective of developing improved technologies. But recognising difficulties in developing technologies for deep water ecosystem, especially when inter-year fluctuations in the water-regime is very high the FSR team was persuaded to respond to research opportunities in different water regimes. Social scientists from Lalit Narain Mishra Institute of Social Science (LNMI) played a pivotal role in refocussing the research areas. Subsequently further farmers pressure led the researchers to explore modifications in land-use pattern during the risk-free environment *i.e.* period between post water receding and onset of flood. The main thrust shifted from varietal screening to agronomical manipulations for effective management of residual moisture in the risk-free season winter or summer.

The initial thrust in the FRS was varietal screening of rice-both released varieties and advanced lines-suitable for lowland, semi-deep and other ecosystems. Most of the 45 entries at the FSR site in 1985 outperformed their counterparts at the experimental station. In 1986 only 22 of the 41 entries taken to the *chaur* survived. The next 2 years were a complete failure as far as rice was concerned due to erratic monsoon. The significant finding of the varietal improvement programme was that none of the varieties tested in the on-farm trials could out yield the local variety *Darmi Jagar*.

The Basket of Technologies

(a) Environmental Characterisation

Environmental characterisation was confined to traditional analysis of soil samples for physical and chemical properties of soil and soil moisture studies at 1/3 and 15 bar.

Agro-ecosystem analysis and other participatory rural appraisal methods were utilised to elicit information from villages to classify land-types into *bheeth* (upland), *uncha chauri* (medium land), *neecha chauri* (low land), and *chaur* (deep water land). Problems associated with different land types and opportunities available were also identified. An excercise in collective mapping of inter-connected *chaurs* resulted in mapping of an area of 40 sq. kilometers. Villagers drew on their detailed local knowledge to identify and explain problems arising out of the complex interlinked hydrology of a chauri or chaur. It also highlighted the limitations of technological innovation in absence of appropriate institutional innovations to mobilise collective actions and manage resource with common property resource characteristics.

(b) Varietal Screening

Exploratory trials with released rice varieties and screening of advanced lines of deep water rice were entry point. Subsequently, varieties were tested under shallow, semi deep and deep water rice eco-systems for their tolerance to drought, water logging, complete submergence, elongation ability, photoperiod sensitivity, seed dormancy and resistance to insect pests. Large number of advanced lines for various land types were also screened.

Farmers found it difficult to manage small lot of seeds of a large

Plate 4 : Rice Varietal Trial

number of advanced lines. Further in deep water context it was difficult to maintain the identification of different lines due to floating and lodging. Farmers showed keen interest in soup breeding (*Ghatra*) trials. Seeds of advanced lines were mixed and sown in deep water and semi-deep water regimes. Plant surviving natural selection pressure were identified and seeds collected for line multiplication at experimental station. These lines were then brought back to the FSR sites for further evaluation.

(c) Agronomical trials-Exploiting risk-free environment

Trials on very early maturing varieties of rice (to escape flood) and boro rice (dry season rice) to exploit residual moisture were laid out. Initial exploration of the farmers had been very encouraging. However the latter had given clues to researchers to initiate trials on nursery management during winter season, weed control and nutrient management. Summer rice trials (very early duration varieties) also indicated that farmers with their precise understanding of agro-climatic variables could easily reduce dependence on irrigation by synchronising nursery raising and transplanting in *Rohini nakshatra* (pre-monsoon period) thereby capitalising pre-monsoon showers.

Farmers reasons for prefering early rice varieties (110-120 days) were

(i) Search of alternate crop in risk free season.

(ii) Direct seeding of summer rice to enable farmers raise a crop when no other crop could be grown and

(iii) Harvest from summer rice is available at a time when majority of households have depleted food grain stock.

(d) Nutrient management

(i) Basal dose of nutrients in deep water trials

In deep water rice eco-system basal application of nutrient was tried to fasten the vegetative growth of rice plants before the onset of monsoon. Rice plants with elongation capacity then have better chances for surving under submergence. However since 1985 Talgarha *chaur* experienced flash floods and prolonged water stagnation. Except 1989 none of the deep water rice experiments could survive. 1989 was good year for the *chaur* rice crop due to low rainfall. Efficacy of this treatment under such situation could not be established.

(ii) Urea super granules vs prilled urea trials

1989 was the only year when this trial survived . Farmers questioned utility of testing an input on their plot which was not available commercially and easily.

(e) Mixed cropping trials

Chaurs have been indegenously classified by local farmers into *Lakshman* (shallow lowland), *Bastar* (medium lowland) and *Darmi*

Jagar (deep water) based on land races of traditionally grown rices (Table - 1). This classification also approximates the water receding pattern of *chaur*; which influences date of sowing of different crops. However, precipitation pattern in the previous year is also a major determinant of water-receding pattern, determining crop establishment.

Lakshman land type is used for raising a mixed crop of mungbean, sorghum and rice. *Bastar* land type is usually utilised for raising a mixed crop of mungbean-sorghum-rice,while *Darmi-Jagar* land type is utilised for seeding rice in flooded field where from June-October water level increases and subsequently recedes in December when the crop is harvested.

Various mixed cropping trials were conducted in different land types. Varietal screening of mungbean in a mixed cropping context revealed that local varieties outperformed released varieties like K 851 and Sona. Rice + til (oilseed) were found to be the most profitable combination at experiment station. However at *Talgarha chaur*, performance of *Krishna* (til) + rice did not come up to the expectation. One of the critical factors was water receding pattern and date of sowing.

Sesbania rostrata, S. aculata and Jute performed in well mixed cropping context. Inter-year performance when analysed with rainfall pattern and water regime could yield very important contingent recommendations.

(f) *Rabi* (winter) season trial

(i) Puddled wheat experiment : In order to exploit residual moisture adequately puddled wheat experiment was laid out in medium lands. Farmers appreciated this trial but yields were not according to the expectation. There is scope for further improvement in the technology.

(ii) Verietal screening of wheat : Several advanced promising lines were screened and few lines with better performance under chaur condition were identified. HP 1490 was most promising. Its seed was multiplied and disseminated through farmers-to-farmer diffusion process.

Plate 5 : Mixed Cropping Trial

(iii) Screening of oilseeds varieties : Several varieties of rape seed were tested. *Varuna* was found to be best among the lot. Farmers adopted *Varuna* variety which spread through farmer to farmer.

(g) Unpredictable water regime

Researchers had not reckoned with the unpredictable behaviour of water in the interconnected chaur. For instance, one varietal screening plot in *Berai chaur* had a water depth of 20 cm on 31 July 1987 and 60 cm on 12 August. Flash-flooding during the night damaged the crop. In the morning of 13 August 1987, the water depth was 160 cm, and the water level kept rising thus creating conditions like flashflooding, complete submergence and protracted water stagnation (6-8 months). Impeded drainage conditions as prevelent in *chaurs* can not be simulated in the experiment station. As a result, none of the research station promising lines could survive the conditions of the *chaur*.

Fluctuations in the water level of chaur, both in a year and over the years, depend on total precipitation received, its distribution, water inflow from the upstream *chaur* and the nature of impediments and congestion in the drainage course downstream (Fig. 6a to 6d).

Initially, agronomic trials merely replicated conventional experiments at research stations. However, they were either modified or discontinued considering farmers' feedback. For instance, experiments on urea supergranules, efficiency of nitrogen in *deepwater* rice (DWR), and top dressing of fertilizers were dropped. Weed control in DWR could never be implemented.

(h) Redesigning Experiment on Fish Culture in *Chaur*

Traditionally designed few tanks providing free inlet and outlet passage following the natural flow of water for capture fisheries are being practised in the *chaur* for long time. The farmers have their own methodology and system for fishing in *chaur*. Rice-fish culture is an age-old practice but the productivity of both rice and fish is extremely low. Presence of tanks in the deepest portion of *chaur*, have given the opportunity to shift from capture fisheries to culture fish farming with higher and assured production.

With a view to generate information on culture fisheries in *chaur* a participatory experiment was designed and conducted during 1991 in four tanks owned and managed by fishermen of Talgarha village (Kumar *et. al.* 1995). In all the tanks fingerlings comprising of six major (Indian & exotic) carps were introduced and inlets-outlets of these tanks were barricaded with *"chilaun"* made of local materials, knowledge and skill to check the exit of fishes but to allow the running water in the tanks.

It has been observed during the experimentation that all participatory famers except one initially adhere to the jointly prepared experimental lay out but they made certain changes in the mid way of the experiment. The changes made by the fishermen resulted in following conclusions :

* The fishermen have unbelievable level of wisdom and experience to apply befitting changes in their decision according to space and time for minimising the risk.
* They are well conversant with the impact of micro and macro environments on fish farming in *chaur.*
* The local fishermen posses vision to analyse with precision and forecast the variable environment, influencing the fish farming in *chaur.*
* The level of risk bearing capacity is not identical rather it differs from fisherman to fisherman and primarily influenced by the economic status of the individual fisherman.
* The nature of ownership of the tank was noted to be important factor to guide the decisions.

At the end of the experiment the production of fish was recorded for quantity and income obtained from the tank. The results surprised the researchers because the tank followed the original plan of lay out and was expected to give higher fish production gave significantly low production and extremely low income as compared to others. The find out the reasons of low fish production in the experimental tanks, the whole issue was put to farmers analysis. Farmers traced the cause that delay in timing of barricading provided entry of wild carnivorous fish along with *chaur* water, negated the expected result.

Farmers suggested following modifications in the approach to lay out the fish experiment in *chaur* :

- In off season i.e. before the start of rainy season, the tank should be thoroughly cleaned and barricaded to check the entry of wild fishes in the tank.

- Regular monitoring of the growth of fishes and quality ... water should be done.
- The fingerlings collected from the river Ganges will be used as it is suitable for *chaur.*
- Planning should be done well in advance to check the overflow of water from the tank.
- The fingerlings of cutlure fishes be introduced in the tank in the month of July or early if situation allow.
- Fish harvesting be done as late as the situation permits to allow longer growth period.

The recommendations provided by the farmers were tested and found profitable and suitable as a low external input technology for the *chaur* farming system.

4.6 Alternate land-use strategies

Farmers have shown their preference for mixed cropping trials. Even in years when rice could not be harvested, farmers could harvest companion crops such as sorghum and jute (in 1987) and mungbean and sorghum (in 1988).

But the most striking chance being attempted by farmers in the *chaur* is incorporation of *Sesbania rostrata* in their mixed cropping systems. *S. rostrata* was taken to the chaur in an exploratory trial in 1988. Farmers were initially reluctant to adopt it, but were surprised to find that *S. rostrata* could survive flash-floods and submergence. Neither land races of rice, which were well adapted to the chaur conditions nor entries introduced by researchers could survive flash floods, submergence and protracted water logging. Farmers could not use *S. rostrata* for green manuring because of waterlogging. However, they found the dried stems a good fuelwood. Continued flooding for 3 years had made fuelwood very expensive, *S. rostrata* appeared in the nich of time as a substitute.

In 1989, preference for *S. rostràta* was wide-spread. With the help of willing farmers, several possibilities of incorporating *S. rostrata* were explored. Never-the-less. *S. rostrata* alone, rice + S. rostrata, and rice + mungbean + *S. rostrata* were tried on different land types. Farmers of the chaur and the FSR agronomist monitored their performance. Subsequently, a well designed exercise was carried out to elicit farmers' preferences and evaluation criteria.

In *Darmi-jagar* land type, the vegetative growth of *S. rostrata* was so profuse that it stunted the growth of other companion crops. In *bastar* land type, farmers could get multiple pickings of mung and harvest sorghum as fodder. They also harvested good yield of rice during the year from the *chaur*. *S. rostrata* came in handy as fuelwood in post harvest processing (parboiling). Farmers continued to devise alternative uses for *S. rostrata* : leaves in the initial stage as cattle feed; sticks for thatching roofs, and for constructing temporary barricades in the *chaur* both as a filtering device and to prevent fish from leaving through water streams.

The farmers gave several suggestions for planning the trials in 1990. According to them the ideal niche for *S. rostrata* is bastar area

Plate 6 : S. rostrata in Deep water condition

Instead of traditional rice + mungbean + sorghum mixed cropping, they suggested an experiment with mungbean + *S. rostrata* and sorghum + *S. rostrata.*

In Darmi- Jagar land type, the farmers did not want to discontinue *S. rostrata.* In 1990, as per their suggestions, two treatments of rice + *S. rostrata* were tried : (i) thinning *S. rostrata* to regulate plant density for working out optimal row and plant spacing and (ii) staggered planting of *S. rostrata* to ensure an appropriate canopy arrangement that would allow rice plants to grow.

4. 7 Farmers' experimentation

Rice is the staple diet of households around the *chaur*. However, a series of unprecedented floods in three consecutive years beginning in 1986, depleted the stocks of grains, and even well-to-do households were put to severe hardships. Farmers looked for alternative food-grain crops during the risk-free season.

In the winter of 1988, puddled wheat was tried, using the receding water. Puddled wheat concept had been transferred from another FSR project in eastern Uttar Pradesh. The Ford Foundation consultant Dr. K. C. John while visiting the *chaur*, broached the concept to farmers and agronomists. Farmers readily accepted the idea (John 1988). An innovative tenant farmers, Mr. Lakhan Sahni, who had always wanted to try puddle wheat, immediately laid out his own experiments. His yield was higher than that of the scientist-designed and scientist-managed trial. Immediately after sowing the chaur received an unexpected spell of rain. Mr. Sahni drained his plot manually that could not be done on the experimental plot. Anticipating the possibility of the seed being eaten by birds, Sahni treated the seeds with a concoction of BHC powder and kerosene. Consequently, plant density were high due to protection from the birds on his plot (John 1989a). Many farmers evinced great interest in continuing puddle wheat experimentation with major refinements in the next winter. Though similar conditions did not occur in the subsequent year, the area under wheat (normal condition) has increased many times as a result of farmer-to-farmer diffusion of wheat seed (variety-HP 1490).

The FSR team decided to embark on varietal screening to find wheat varieties suitable for different niches. Several research questions were also identified based on the farmers' experience with puddle wheat and the researchers' observations. Many of the questions could be explored only under the highly controlled conditions of the experiment station (John 1989b). The puddle wheat experiment did not satisfy traditional wheat researchers, nor was it included in the FSR's approved technical programme to find a suitable method of sowing and optimum seed rate in the *chaur*. But wheat became an important component of farming systems in the chaur.

4.8 Linking formal research with informal experimentation

Another major initiative in finding alternative food crops in the *chaur* was summer maize. The FSR team was surprised to find large areas under summer maize. Many of the farmers experimenting with summer maize were tenant farmers. In the first year they wanted their experiment to succeed and invested heavily in buying water from their landlords. The crop succeeded. But crop sharing on a 50 : 50 basis did not encourage resource—poor farmers to continue with the experiment.

Taking the clue from the farmers' new experiment in summer maize, the FSR agronomist intensified his search for very early rice varieties (60-90 days duration) and summer rice varieties (end March-beginning April) suitable for *chaur*.

Very early varieties of rice were seeded directly. Most of the collaborating farmers ignored the exploratory trial in the early phase. This was the first time that such a crop was screened in *chaur* conditions. Only when the crop flowered did the farmers begin to show interest in its management. Low seedling vigor, low tillering and smaller panicles were the other reasons for neglect. Farmers showed distinct perference for yield and plant type of Pusa 2-21 and CR 44-35 which matured in 120-125 days. The reasons for continuing the direct-seeded rice trial (John 1990) were as follows :

1. There is need for an alternative to the main rice crop in the risk-free season.
2. Direct seeding of summer rice (end March-beginning April) enables farmers to raise a crop when no other crop can be sown.
3. Harvest from direct-seeded rice coincides with the period

(mid-July) when food grain stocks in most households are low.

In the local land classification, *lakshman* land type has the potential for direct-seeded summer rice. The farmers believe that under normal circumstances, the upper fringes of the *chaur* where water recedes by the first week of January could be planted to wheat. Except for this region, other parts of the *chaur* could be planted to 110 to 120 days rice at the latest by first week of April. Land available from the second week of April is traditionally devoted to a mixed cropping of rice + mungbean + sorghum. Several farmers' suggestions to refine the treatments have been incorporated in the new set of experiments. They include evolving agronomic techniques such as plowing, change in the date of sowing, weed control, nutrition management and supplemental irrigation (John 1990). The farmers around the *chaur* are keen to exploit the risk-free period and maximize the production potentials of their small landholding. In a good year, farmers could obtain yields as high as 4.0 t/ha (darmi-jagar) in the deepwater rice ecosystem. In contrast, summer rice experiments in farmers' fields yielded 5 to 6 t/ha.

FSR researchers in the *chaur* had also become sensitive to local parameters of predicting the weather. However, the cost of this change was borne by a resource poor tenant farmer, Govind Ram. Summer rice (transplanted) was also tried initially. Pre-experimental briefing emphasized sowing the nursery by end of March and transplanting seedlings within 30 days. Govind Ram adhered to the pre-experimental briefing (nursery sowing on 24 March and transplanting on 20 April 1989). The subsequent water requirement was very high. Initially, he purchased water to irrigate summer rice, but could not afford it for long. The continued dry spell and inability to buy water for irrigation compelled him to reduce his commitment to the experiment. The experiment failed.

However, a similar experiment in summer rice (transplanted) was success in Ramdeo's-a tenant farmer and Baliram's plots. In spite of persistent advice, they delayed raising the nursery by more than a month. The experimenter was discouraged when the farmers ignored the recommendations and chose to transplant summer rice with the emergence of *Rohini* (early monsoon period) believed to be a precursor of premonsoon showers. Two farmers irrigated the land and puddled it for transplanting summer rice in late May to coincide with *Rohini*. Frequent rains in the premonsoon period helped them and reduced dependence on irrigation. After transplanting both Ramdeo and Baliram had no occasion to irrigate the summer rice. Under low management conditions, they harvested 2.4 t rice/ha in July (John 1990). Farmers' experiences and suggestions have led to modifications in experimental treatments.

Research opportunities identified in traditional fish culture and fishing practices, integrating smaller ruminants with existing cropping systems, alternative uses of crabs and snails as poultry feed, etc. (Saran-1989) could not be exploited. The crop oriented thinking of the research institution and lack of inter-institutional collaboration were the major reasons for the failure.

SUMMARY

Farming Systems Research programme began with the development of technology for rainfed lowland cum deep water rice ecosystem. High quality seeds from advance lines and proven agronomic practices were tested in On-farm trials. Fluctuations in the on-set of monsoon, flash flooding, water stagnation due to impediments in natural drainage and irregular water receding patterns were some of the reasons for the failure of promising technologies. The limited range of search and quasi resolution of problems compelled researchers to dialogue intensively with farmers. It soon became apparent that varietal screening and agronomic practices were too narrow a focus to have any significant impact on the livelihood of farmers.

This chapter, based on experience and insights gained working with farmers as partners in technology generation presents a method of balancing the component and systems technology development process. The method has been found effective in developing an integrated farming system in flood prone regions. It is argued that regular monitoring of farmers risk-cropping mechanisms and elucidating the underlying principles and role of risk allocation, will enable the FSR team to adopt a gradual and incremental approach to evolving a portfolio of enterprises strategy to diversify enterprises, which lower systematic risk and allow risk-free seasons and sectors to be exploited.

It is recommended that in a complex dynamic and risk-prone eco-system FSR scientists investigate ecological based land-use systems of farming community rather than being circumscribed by entry point strategy. Adapting an integrated land and water resource management perspective has opened up new and interesting lines of enquiry. It also encouraged a search for organic linkages among crop livestock-fisheries-horitcultural enterprises in flood-prone region. Implications for methodological innovations have been explored.

REFERENCES

Ashby, J. A. (1986). Methodology for the Participation of small Farmers in the Design of On-farm Trials Agricultural Administration and Extension 22 pp. 1-9.

Ashby, J. A. and D. H. Pachico, 1989. Farmer evaluations of technology. A handbook. IPRA Projects. Centro International de Agricultura Tropical, Cali, Colombia. 118 p.

Attah-Krah, A. N. and P. A. Francies, 1987. The role of On-farm trials in the evaluation of composite technologies : Alley farming in Southern Nigeria. Agric. Syst. 23 : 133-152.

Biggs, S.D., 1980. Informal R & D Ceres 13 : 23-26.

Biggs, S. D. and Farington, J. (1989). "Social Science Analysis in Agricultural Research : A review and conceptual Framework IDRC Ottawa.

CWRS (Centre for Water Resources Studies), 1990. Assessment of wasteland problems in humid alluvial regions-a case study in North Bihar. Bihar College of Engineering, Patna University, India.

Garrity, D. P. (1990). Choice in On-farm experiments. Researcher Implemented, or Farmer Participatory Trials". Paper presented at the Workshop on On-farm Research Design, NDUAT Faizabad 13-17 February 1990.

Hilderbrand and Poey (1985). "On-farm trials in Farming System Research and Extension". Lynne Rienwer Publisher Inc. Boulder Colorado.

John, K.C. 1988. Suggestions for utilizing receding water in the chaur during rabi season (Mimeo).

John, K. C. 1989a. A narrative report on farming systems research project at Talgarha chaur (Mimeo).

John, K. C. 1989b. Strengthening farming systems research project at Talgarha chaur (Mimeo).

John, K. C. 1989c. Indigenous environment classification system in semiarid region of Gujarat. A report prepared for Aga Khan Foundation, Geneva.

John, K. C. 1990. Farmers' evaluation of farming systems research experiments in Talgarha Chaur (Mimeo).

John, K. C (1990). Design and analysis of On-farm experiment. Issues in Research Methodologies". Paper presented at ICAR-Ford Foundation Workshop on Farming System Research in Eastern India, India International Centre, New Delhi 16 July, 1990.

Kumar, A; K. C. John and S. Saran. 1990. Sustainable agricultural development in chaur : Linkages among policies, transactions and ecological risk. Paper presented at the Workshop on Linking Farmer Participatory Research to Planning Implementation. 6-9 March 1990, University of Kalyani.

Kumar, A; Sinha, M. M. and Ahmad, S. H. (1995). Farmer oriented analysis for redesigning experiment on fish culture in *chaurs* of Bihar. Fishing chimes 15 (4) : 27-29.

Menz, K. 1980. Unit farms and farming systems research : The IITA experience. Agric. Syst. 6 : 45-51.

Okali, C. and J. E. Sumberg. 1986. Examining divergent strategies in farming systems research. Agric. Admin. 22 : 233-253.

RAU (Rajendra Agricultural University). 1989. Technical programme rice. University Research Council. Rajendra Agricultural University, Pusa, India (Mimeo).

RAU (Rajendra Agricultural University) 1990. Technical programme rice, University Research Council, Rajendra Agricultural University, Pusa, India (Mimeo).

Rangasamy, A; V.S. Sanmuga Sundaram; M. Subbarayalu and S. Sankaran, 1989. Integrated farming systems management-a viable approach. Centre for soil and Crop Management Studies, Tamil Nadu Agricultural University, Coimbatore, India (Mimeo).

Richards, P. 1985. Indigenous agricultural revolution. Hutchison, London, and Westview Press, Boulder, Colorado.

Saran, S; P. Mishra; A. Kumar; and A. Salman, 1992. "Farming System Research Approach in Risky Agro-eco-systems of north Bihar, India". Journal for Farming System Research, Vol. 3 No. 1 University of Arizona, Tucson, U. S. A.

Saran, S; K. C. John; A. Kumar; A Salman; M. M. Sinha and P. Mishra 1993. "Integrated Farming System Methodology for High-Risk Ecological Zones". Journal of Asian Farming System Association, Vol. 1 No. 4 pp. 463-477 ATT, GPO Box 2754 Bankgkok 10501.

Thakur, R; A. K. Singh; R. Singh and R. S. Singh, 1990. Survey of deep water rice in Bihar (1986-1989), Rajendra Agricultural University, Pusa, India (Monograph).

Water-Bayer, A. (1989). Trials by scientists and Farmers : Opportunities for co-operation in Ecofarming Research in Kotaschi J (ed) Eco-farming practices for Tropical Small Holidays. Research and Development in Technical Co-operation GTZ Eschobora, pp 161-183.

Chapter – 5

Outstanding Issues

It is being increasingly realised at different levels of development in India that a major part of the answer to the problem of future agricultural growth in the country has to come from eastern India. This premise is supported by the large gap between the actual and potential levels of agriculture development in the region in whatever way the potential is measured. The persistence of such a gap in the face of relatively faster growth in other regions of the country calls for a close analysis of two sets of factors (a) the constraints to the generation and adoption of improved technologies (b) the formulation and implementation of policies that would facilitate the better utilization of available technologies.

The challenge is specially great in the rainfed (flood prone and/or drought prone) areas which still represent about 70 per cent of the region. The generation of improved technologies for these areas has proved very difficult through the use of Conventional Transfer of Technology (TOT) approaches to agricultural research and extension and on the hypothesis that TOT approach is inappropriate to meeting the needs of these very diverse risk-prone conditions. Scientists in Bihar and eastern India have in recent years been looking for new ways to generate technology through participatory research with farmers.

One of the advantages of Farmer Participatory Research (FPR) is that it enables scientists to understand in detail the complex socio-economic and agro-ecological factors that influence farm families decision making in rainfed areas in a way that it is impossible through on-station research, alone. By interacting with farmers as partners in research, scientists can be greatly helped in identifying promising new lines of technology improvement and through participatory processes of social/environmental mapping larger recommendation domains can be identified.

Policy Implications

As has been argued above, research into improved technologies for diverse agro-eco-systems in Bihar and eastern India needs to be complemented and reinforced by research into appropriate agricultural policies. Like agricultural research the FPR work has so far tended to accept the prevailing policy environment setting the boundaries within which technology options are to be explored. Policy analysis on the other hand has tended to be largely top-down process.

The need of the hour is to identify concrete measures that would address the following issues.

How can the FPR be expanded so as to (a) identify more effectively the impact of policies on different categories of farm families in different agro-eco-systems (b) feed back insight from the field that could be used to reformulate policies and (c) how can the processes of policy planning and implementation at local, state, and national levels be adapted to draw upon new information obtained through FPR and other related activities.

On the basis of our FPR experience in the risky agro-eco-systems in Vaishali district we have attempted here to throw light on policy gaps in the area of agricultural research and institutional support to link trade, agro-processing and credit with technological interventions. This excercise helped in identifying areas of initiating dialogue between policy makers and farmers.

5.1 Interconnected Flood Prone Basins (*chaur*) of North Bihar

Northern Bihar is interspersed with many *chaurs*. It is estimated that about 0.6 m. ha. land are under chaurs (CWRS, 1990; Thaiur *et al,* 1990). They are often interconnected. The interlinked hydrology of a series of *chaurs* gives rise to a very complex, high-risk and flood-prone ecosystem. Frequent flooding and inundation brings along with it high fertile soil deposits. However, flash floods and submergence during the monsoon period and protracted water logging create unfavourable conditions for cultivation. The FSR *chaurs* fall in the midstream of a natural drainage course running from north to south between a tributary of the Gandak River and the Ganges. Villagers around the *chaurs* drew on their detailed knowledge about the hydrology of the *chaurs* and the nature and extent of the problems. They exhibited their capacity to address constructively issues relating to the management of the overall water regime. A collective exercise in mapping with the help of villagers around these *chaurs* resulted in drawing of an area spanning about 40 Km having 31 interconnected *chaurs*. Besides, they also defined precisely the sources of different problems and the casual linkages involved. Solutions to several of the problems would invite negotiations at different levels of collectivity (Bottrall and John 1990).

Low and uncertain productivity of the interconnected *chaurs* was also analyzed in a farmer participatory mode. A decision tree tracing the casual linkages is shown in Fig. 10. Submergence and protracted waterlogging are important deterrent to intensive agriculture in the *chaur*. Submergence is positively associated to water inflow into the *chaur* and precipitation received in the catchment.

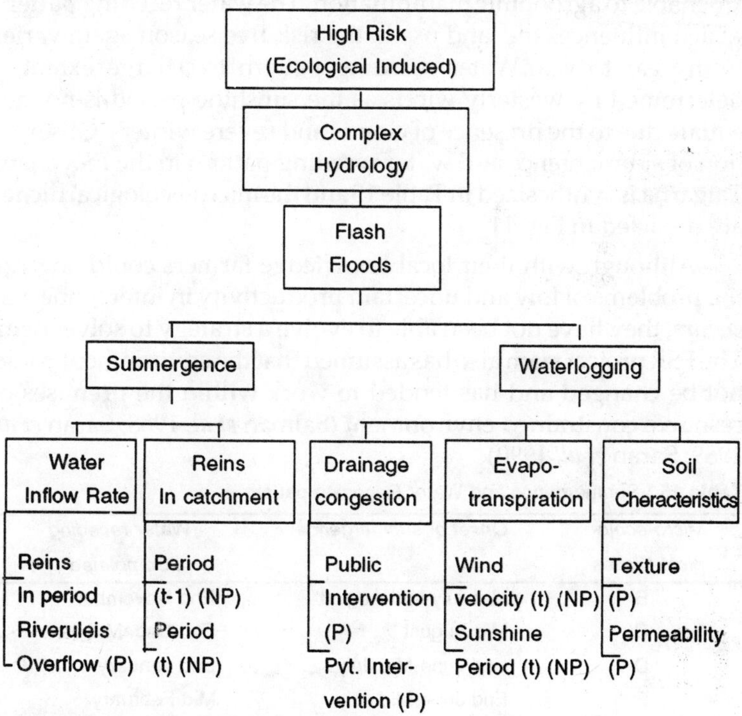

(P) (NP) Policy and Non Policy Variable.

Fig. 10: Low and Uncertain Productivity in interconnected *chaurs*

One of the major contributing factors of waterlogging is the congestion in the natural drainage course. Construction of embankments, canals, and roads at the public sector initiative but without proper drainage lead to the drainage congestion of the inter-connected chaur (CWRS 1990).

In the collective diagnostic analyses, farmers were able to identify not only major sources of congestion as embankments and roads but also incomplete drainage and faulty implementation of drainage scheme by the water resources bureaucracy. Private interventions in several downstream *chaurs* such as blockading of natural outlets, construction of embanked ponds for fishery raising, *etc.* were also identified.

The physico-chemical properties of the soil also do not allow vertical drainage of water. However, in the long-run these are amenable to agronomic manipulation. The water receding pattern which influences the land use in the risk-free season again varies from year to year.Water receding pattern to a large extent is determined by westerly winds as the sunshine period is not adequate due to the presence of clouds and severe winters. Observation of submergence and water receding pattern in the FSR chaur Talgarha is synthesized in Table 15 and the micro-ecological niches are depicted in Fig. 11.

Although with their local knowledge farmers could analyze the problems of low and uncertain productivity in interconnected *chaurs*, they have not been able to evolve a strategy to solve them. The FSR project team also has assumed that the environment could not be changed and has tended to work within the premises of resource constrained environment (Salman *et al*, 1988; Saran *et al*, 1989; Saran *et al*, 1990).

Table 15 : Submergence and Water Receding pattern

Micro-ecological Niches	*Onset of submergence*	*Water receding Completed*
B	Mid July-Mid August	End November
C	Mid August	End Dec.-Mid Jan.
D	End June-Mid July	End January
E	End June-Mid July	Mid February
F	End June-Mid July	Mid Feb. Early March
G	End June-Mid July	Mid-End March
H	End June-Mid July	Mid-End April

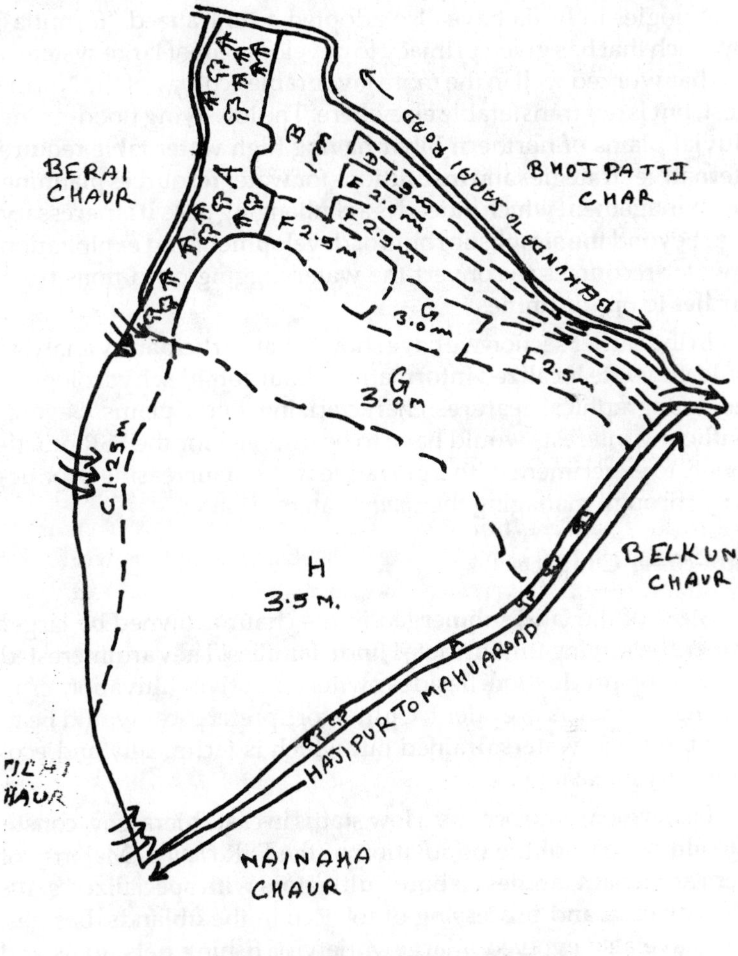

Fig. 11 : Water Regime Map of Talgarha *Chaur*

5.2 Sustainable Development of the *Chaur*

We have attempted to draw attention to the inadequacy of technological innovations in isolation and the need for establishing congruence between broader policy regime and the biophysical environment of rainfed agriculture in the flood-prone basins.

Water resource management issues in the *chaur*.

As with the Green Revolution technologies, water resource use technologies in India have also adopted a centralized "formula" approach that has given primacy to development of large systems. This has worked well in the more favourable ecology of the North-West, but is not transferable elsewhere. The low-lying flood-prone alluvial plains of northern Bihar having high water table require alternative strategies and institutions for water resources planning and management which indeed is a challenging task. It is necessary to go beyond the simplistic route of development and exploitation of water resources to convert the water logging conditions from hurdles to opportunities.

In the earlier section we have shown that participatory analysis could generate localized information about complex hydrological and topographical features characterizing flood plains. Several conflicting interests would have to be resolved for the FSR practitioners to experiment with innovative ways of increasing productivity through managing the *chaur* waters.

Intra-*chaur* Conflicts

Most of the land submerged in the chaur is owned by larger farmers belonging to high caste Hindu families. They are interested only in crop production. Besides owner operative cultivation, crop sharing tenancy is prevalent. Cultivators preferences would be to have the flood waters drained out which is technically and economically infeasible.

Fishermen, who occupy a low status in caste hierarchy, constitute almost half of the population of the FSR *chaurs*. Majority of them are tenant landless labour cultivators with specialized skills in cultivation and processing of tobacco in the uplands. Besides, they have also evolved a large variety of fishing nets, tools and practices to harvest fish in different months depending on the water depth and fish species. The fish enter from the upstream *chaur* during the flood season and capture fishing is predominant.

Past observations have indicated substantial inter-year fluctuations. When water levels are high as in 1988, it is a good for fish and a bad for crops. When they are low, as in 1989, the situation is reversed. The fishermen have controlled their environment in multiple ways by closing the sluice gates at the end of Talgarha chaur for retaining more water for a longer period and thus increasing the likelihood of larger fish harvests. However, this also leads to protests from farmers dependent on crop harvest for eking out their livelihood within the *chaur* and further upstream.

Fishermen have also tried putting nets across the outlet of the *chaur* and stocking it with additional fish. But area around the outlet is controlled by people from the downstream village and the risk of losing the investment is very high.

In the most depressed portion of the *chaur*, some of the fishermen families have dug ponds on their own lands for impounding the receding water during the months of April / May. Fishing rights are restricted to the pond owner during that period when large amount of fish could be captured and sold at relatively higher prices. The rest of the fishing families who have no ponds have virtually open access to fisheries anywhere in the *chaur* except during April / May period. However, some of the pond owning fisherman families control the fish market and prices.

Villagers have shown their enthusiasm in converting their capture fishery to fish culture system. But the choice of techniques would have to be influenced by a clear understanding of its implications of property rights and ownership (Bottrall and John, 1990).

Inter-*chaur* Conflicts

During the collective mapping exercise, farmers were able to define precisely the sources of drainage congestions along the natural drainage course. Besides fishermen-community's interventions in backing-up of water, several private embankments were also identified in the downstream chaurs for regulating the inflow of flood. Instance of inter-*chaur* conflicts, which often turned violent, were narrated by villagers. Though they have shown their capacity to think knowledgeably and constructively about issues relating to the management of the overall water regime, they would require assistance from external development agents for evolving a strategy for resolving inter-*chaur* conflicts.

Conflicting interest among development bureaucracies and local community

The most immediately significant development bureaucracies for the farmers around the *chaur* are Departments of Irrigation and

Agriculture. The FSR project site is within the jurisdiction of Gandak Command Area Development Authority (GCADA) under its Tirhut Canal Division. The GCADA is compartmentalised into Canals, Drainage and Flood Control Section-each section works in an insulated manner and lack inter-sectional communication and co-ordination. As a result, the canal section which is responsible for construction and maintenance of the main canal, the distributaries and the minor canals, is planning to extend a minor canal to the Talgarha *chaur*. The farmers of the region have petitioned against the construction of a minor canal. Their problem is drainage rather than irrigation. Discussions with the Executive Engineer of the Tirhut Canal revealed that they had assumed the existence and smooth functioning of a drainage system. They had also assumed that the Department of Agricultural responsible for the construction and the maintenance of field channels-has discharged their responsibilities.

The farmers and the FSR team have yet to evolve a development strategy which could contribute significantly to enlarging the range of production options open to farmers around the *chaur*. However, the FSR practitioners have begun to design a series of policy dialogues with the development bureaucracy to find solutions to the issues of management of overall water regime. Problems associated with collective action and decision making and their treatment through the FSR approach are discussed elsewhere (Bottrall and John, 1990).

Developing Water Markets

Vulnerable groups of households, besides supplying agricultural and domestic labour, are tenant farmers. A series of unprecedented floods in 3 consecutive years beginning in 1986 had depleted their foodgrain inventory. In a rice-dominant monocropped area, farmers have been looking out for alternate foodgrain crops which could be grown during risk-free season in upland and lowland areas. The FSR team was taken by surprise in 1988 to find a large area under summer maize. Most of the growers experimenting with summer maize. Most of the growers experimenting with summer maize were tenant farmers. In the first year they wanted their experiment to succeed and invested heavily in buying water from few landlords who owned water extraction machines. The crop succeeded. But tenurial pattern, viz., crop sharing without cost sharing, discouraged resource-poor farmers to continue with the experiment.

Picking clues from the farmers' new experimentation, the FSR team intensified its search to find very early varieties of rice which could be grown as a substitute to summer maize. Very early rice varieties have several advantages, such as lower water and labour requirement. Three sets (60, 69 and 90 days duration) of rice culture were put on on-farm trials. Direct seeded rice could be grown in large area provided life-saving irrigation is ensured. A few ingenious farmers saved their transplanted rice by delaying nursery raising and transplanting to coincide with pre-monsoon showers. If a network of shallow tubewells is established, it could bring in radical changes in cropping systems and intensities, besides providing an outlet for vertical drainage in the chaurs. In similar socio-economic and agrarian settings in the central alluvial plans of West Bengal, the silent groundwater revolution, besides changing cropping systems, has also developed a highly competitive water market. In such a case of competitive and efficient water market, it has been hypothesized that water buyers might benefit almost as much as water sellers. The development of water markets will call for conductive government policies which treats drainage of water as a critical element in the overall water resource management, good access to institutional credit and investment in evolving attractive but appropriate water extraction technologies.

5.3 Public Policies and Programme

The FSR project site comes under the jurisdiction of Gandak Canal Command Area Development Authority (GCADA) under its Tirhut Canal Division (Hajipur). GCADA is organised similar to its parent organization, *i.e.* the Department of Irrigation, Government of Bihar into :

1. Canals Section
2. Drainage Section
3. Flood Control Section

The Canal Section is responsible for construction and maintenance of the main, the distributaries and the minor canals while the construction and the maintenance of field channels are entrusted to the user department the Department of Agriculture. Sarifabad minor canal (SMC) of Mallikpur Branch Canal (MBC) is planned to end in Talgarha *chaur*. Farmers of the region have petitioned against the construction of minor canals. Their problem is drainage rather than irrigation. SMC has a command area of 1500 acre. During discussions with the Executive Engineer, Tirhut Canal (EE/TC), Hajipur these issues were raised. EE/TC was emphatic that the Canal Section is mandated to create irrigation facilities in this area. They have assumed that the drainage system is already existing and operating smoothly. They also assume that the Department of Agriculture has already created the field channels.

The Government of India supports the drainage work, though the state is responsible for proper functioning of drainage schemes. The villagers had identified incomplete drainage construction activities. Executive Engineer, Drainage Section (EE/DS) corroborated that in 1979, the erstwhile Department of Water Resource Management had initiated a drainage scheme for interconnected chaurs. It envisaged deepening of the Ghaghra and leading the drainage along the course of Ghaghra to the Ganges. Due to changed priorities and transfer of offices, work came to halt in 1985. Since November 1989, the scheme is reactivated. The length of drainage system is expected to be 188,000 ft. long. Survey will be recommended after the water recedes in the chaur, i.e. November 1990.

The major stumbling block in Bihar and Eastern India in general has been the adherence of State Irrigation Departments to a "North Western" model of development that has given primacy to development of large canal systems. As with the Green Revolution technologies, water resource use technologies have also adopted the centralized "formula" approach, which happened to work well in the more favourable-ecology of the Northwest, but is not transferable elsewhere. The lowlying, flood-prone, high water-table areas such as the present one require alternative strategies and institutions for water resources planning and management, which indeed is a challenging task. Sufficient independent information about complex hydrological and topographical features characterizing flood plains is lacking. FSR, AEA and PRA techniques could contribute towards participatory analysis and generation of localized information. It is necessary to go beyond the existing route of development and exploitation of water resources to convert the waterlogging conditions from a hurdle to apportunities.

The FSR team, besides designing a series of policy dialogues have also started exploring possibilities of a collective fisheries improvement programme.

5.4 Diversified Portfolio of Enterprises

Sustainable development of high-risk ecosystems could be ensured by increasing the diversity. Farmers of the region have evolved a mixed cropping system such as rice-mungbean-sorghum and different sowing and crop establishment methods for hedging against risk. In its attempt to develop an integrated farming system, the FSR team had been exploring integration of crops like sugarcane, vegetables, flowers and non conventional crops in the *chaur*. (Kumar *et. al.* 1990)

Sugarcane Cultivation in *Chaur* Condition

Sugarcane was traditionally grown in rotation with deepwater rice in the chaur. A historical profile of sugarcane cultivation, its discontinuance and reemergence is traced in Table 16. Due to irregular payment, a long waiting line before the factory gate opens to weigh the sugarcane, and lock up of the sugar factory, Talgarha farmers discontinued sugarcane cultivation around 1980. However, three successive floods in the *chaur* and absence of other alternatives induced few innovative farmers to experiment with BO 91, a variety which could tolerate water submergence for a few days. They depend on private crushing within the village. Reserved area stipulation of the Bihar (Sugarcane Regulation of Supply and Purchase) Act. 37, 1982 clause 31 envisaged that nobody could be allowed to take up private crushing of sugarcane for making gur or khandsari whereas the Guraul Sugar Factory is in no position to encourage sugarcane cultivation in its catchment area.

Two varieties are recommended for submerged area, viz; BO 91 and 109. The sugarcane Research Institute at Pusa does not have plots akin to the *chaur* conditions in the experiment station. It recommendation for management practices for water-submerged area is based on the nonrepresentative plot at the experiment station. Recommendations for nutrient management were also not relevant due to non-comparability of experiment station conditions with *chaur* conditions. Varietal recommendation is not specific for different water regimes. Planting in the *chaur* will depend on the water receding pattern which is also not reflected in the recommendation. Indigenous innovation on date and method of sowing in other chaurs are known. For instance, Upendra Singh, a farmer in Samastipur, has been conducting his own experiments on varietal screening, date of sowing and method of planting. CO 767 was found to have better submergence tolerance relative to recommended varieties BO 91 and 109. Singh has also perfected a planting and basal nutrient application method in standing water (1-2 inches).

Vegetables in the *Chaur*

Farmers in the *chaur* had undertaken tomato cultivation in upland area (*bheeth*) on being motivated by the vegetable and fruit processing plant of the Bihar State Fruit and Vegetable Development Corporation (BSFVDC) in 1985. In 1986, out of about a 25 ha. bheeth land, 17 ha were allocated to tomato cultivation only. Table 17 provides significant milestones of tomato cultivation in the chaur. Once again like sugarcane processing, initial promises disappeared soon. Inefficient management, inadequate logistic support, unremunerative pricing, irregular payment were likewise the culprit this time.

Pulses in the *Chaur*

In Bihar more than 97 per cent of the area under the production of mungbean is in the *chaur* lands. The local variety is generally sown as a mixed crop with deepwater rice and sorghum towards April. Local variety of mungbean, a crop of 60-65 days duration, has small and green grains. It is amenable to a wide range in sowing time, offering multiple picking and has moderate to high resistance to yellow mosaic virus. It has a yield potential of 4-6 qt/ha in mixed cropping system. None of the released varieties of mungbean has outperformed the local variety in the FSR experiments.

Experimental station research is confined to *kharif* (June-October) season whereas *chaur* production is in the summer season. All the research is for raising mungbean as a monocrop. No technologies have so far been worked out under mixed cropping conditions. At the time of the FSR team, local *chaur* mungbean has been taken to the pulse breeding programme.

The Department of Agriculture, under its minikit programme has provided K 851 variety, released for relatively upland conditions. It was observed to be susceptible to YMV and yielding around 1-1.5 qt/ha in mixed cropping system. Discussions with the Block Development Officer of the Mahuwa block revealed that the chaur lands have not been distinguished for distinct agricultural improvement programme. Nor do the BDO office level functionaries receive specific guidelines for the *chaur* related crop technologies.

Research priorities on mungbean in Bihar would have to be shifted from monocrop *kharif* season to mixed crop in summer season. Preliminary dialogues with pulse researchers have been very encouraging. Although there is direct linkage between the SAU Research Unit, many times technologies being disseminated by the State extension machinery may prove to be disastrous. FSR approaches to extension in risk-prone regions would appear a pragmatic policy option.

Non Conventional Crops for the *Chaur*

Failure of the rice crop for 3 consecutive years beginning 1986 had compelled farmers to intensively search for alternative crops. On-farm research on non-conventional crops have revealed feasibility of diversified crops which could be grown at different seasons.

Sesbania, jute and marigold have been integrated in the farming system. Sunflower was outrightly rejected due to absence of marketing and processing facilities. Rajmah (pulse) is still under evaluation (Table 18).

5.5 Lessons for Sustainable Agricultural Development

Cases presented reveal important lessons for sustainable agriculture development in high risk-prone regions. We list salient ones here :

1. Strategy of increased diversity has been favoured by farmers of the chaur. Diversities in crop enterprises, genetic materials, crop establishment methods to suit the needs of different niches could be found in the *chaur*. Diversification has been an important tool to hedge against risk.

However, institutional support and infrastructural facilities-be in the water resource management, processing and marketing of agricultural produce or agricultural research and extension-have not improved risk allocation of farmers in the chaur. Each case highlights policy and programme failure, thus implying that risk had to be managed privately. Farmers have depended on individual ingenuity or close network of family and friends to cope with risk. Such arrangements are quite useful to cope with noncovariant risk. But we have also indicated that yield risks are covariant in the *chaur* due to submergence and waterlogging. Under such a condition, contracts within closed groups which are implicit in nature would not perform well as insurance schemes for yield risks unless they cover extensive geographic area. However, the geographic coverage of closed networks by its very nature would be limited. Consequently, in flood basins, a significant part of the yield risk, with subsistence agriculture, would have to be borne privately. This on diversifying enterprises selection in the *chaur*.

2. Such a risk management strategy has serious implications for agricultural research and extension. Researchers would need to be concerned with many production activities. Whereas the green revolution technologies had focused on monocultures grown under most favourable extended tracts of irrigated area, in risk-prone and marginal areas, research would have to shift from commodity-based to resource-based research strategies. The greater the diversity, complexity and risk-proneness of ecosystems, the more the need to adopt decentralized and resource-based research strategies.

3. The FSR related techniques of AEA and RFA become more critical for agricultural extension workers under risk-prone situations. These techniques not only could make initial modest contribution towards the development of an alternative, farmer participatory, more decentralized and accountable approach to technology generation and dissemination.

4. Farm diversification strategies would increase the transaction costs of trade, transportation information acquisition and input procurement. Policy and programme innovations are required for risk-prone regions that would lower transaction costs and uncertainly concerning a particular risk-sharing scheme

involving a group of individuals. Due to organizational-inefficiency, sugarcane and tomato processing activities failed. These were public sector initiatives but could not become effective in risk management and reduction of transaction costs.

5. This also suggests that the allocation of risk is an integral part of the sustainable development process and should be an explicit part of agricultural development policies.

6. Five policy criteria identified by Conway and Berbier viz; "Political Will", "economic analyses", "appropriate incentives", institutional flexibility" and "complementary infrastructure" become crucial in the creation of a favourable policy environment which would encourage pursuit of long-term development objective such as stability equity and sustainability.

7. Policy dialogue and social partnership among agricultural researchers, the FSR team, NGOs and relevant government sector would go a long way in evolving programmes with conscious concern for reduction of transaction costs and risk spreading.

Institutionalization of FSR

Recognising the advantages of FSR as a vehicle of agricultural development, the modalities of its integration into the mainstream research and extension efforts need evaluation. As on the importance of farming system approach should have the cognizance as a policy programme. What is required is to enunciate the steps needed and the mechanism appropriate for integrating FSR into the national agricultural research and extension systems (Raman *et al* 1992).

It is realized that there is no universally acceptable method for integrating FSR into the mainstream research and extension. Five possible options are available for the purpose (Raman *et al* 1992) (a) conduct of both on-station and on-farm research by the same commodity/discipline oriented researchers with the assistance of a social scientist (b) a separate FSR team to perform on-farm trials on the technology generated at experimental station (c) commodity-discipline oriented researchers backed by a systems Oriented FSR team with area responsibility which identifies the production problems, carries the technology information to the farmers in the influence area and is also responsible for the feed back (d) same as

Table 16 : Historical Profile of Sugarcane Cultivation in Talgarha Chaur

Period	*Significant Milestones*
Historical Reality	Cultivation of Burrokh' or Bhulla' sugarcane in a crop rotation of rice-sugarcane *i.e., darami jagar-burrokh*. Salient features of *Burrokh*. 1. brown, thin but long stem 2. plants and elongation capacity and high plasticity and elasticity. 3. Yield : 100-200 q/ha.
1934	Establishment of Sheetalpur Sugar Factory (SSF) about 40 Km from the FSR *Chaur*. SSF popularized BO 17 and BO 21 and discouraged cultivation of burrokh through crop loan 'Badhani'.
1938	Enforcement of the Bihar-sugarcane (Regulation of Supply and Purchase) Act, 1938.
1942	Farmers of Talgarha *chaur* switch from burrokh to BO 17 and BO 21 which could not be cultivated in deep water conditions. Shift in cropping system of the *chaur* Deep Water regime : *dharami jagar* monocropping. Upper periphery: rice/sugarcane.
1950	SSF plants and machinery shifted from Sheetalpur to Garaul which is 10 Km from Talgarha *chaur*.
1968-1973	Lock up of Garaul Sugarcane Factory due to inefficient management and continued losses.
1974	The Bihar Sugar Corporation took over GSF. Changes in procedure.
1979	Hazipur-Muzzafarpur meter gauge rail line converted to broad gauge but Factory management did not agree to convert line from Garaul Railway Station to factory. Switch to truck-operated collection system. Low reliability. High cost of transportation. Long wait for sugarcane growers at factory gate for weighing of cane.
1980-82	Irregular payment system. Talgarha : Sugarcane cultivation discontinued.
1987	Reduction in number of collection centres.
1988	Garaul factory paralysed due to shortage of sugarcane.
1989	3 successive floods in the *chaur*, no crop harvested. Innovative farmers experimenting with sugarcane variety BO 91 and private crushing.

Table 17 : Profile of Tomato Cultivation in the *Chaur.*

Period		*Significant Milestones*
1982	—	The Bihar State Fruit and Vegetable Development Corporation (BSFVDC) decides to encourage systematic fruit and vegetable cultivation in the State : It envisaged remunerative prices to farmers and value added processing as its instruments.
1985	—	Fruit processing plant commissioned
	—	Tomato growers registered and farmer participatory process initiated by the Managing Director (M.D.) for establishing working relationship, price fixation *etc.*
	—	Informal collection centres organized.
	—	Full payment received by growers at the end of the season.
1986	—	Area under tomato increased to 17 ha in Talgarha *chaur.*
	—	Break-down in trucking facility
	—	Growers suffer losses
	—	Irregular payment
	—	Price rigidity
	—	Partial payment at the end of the season.
1987	—	Collection centre at Talgarha discontinued.
	—	Price differentials between plant and market was Rs. 1 per Kg.
	—	Farmers selling their produce in open market.
	—	M.D. intervenes and agrees for on-the-spot payment at reduced price differential of Rs. 0.50 per Kg.
	—	Farmers willing to support this new initiative.
	—	Deterioration in collection and payment pattern continues.
1988	—	Growers revert to pre-1985 status. Grow tomato for home consumption only.

Table 18 : Non Conventional Crop Enterprises

Crop Potential	*Season*	*Production (in q/ha)*
Jute	Wet	12
Sesbania rostrata	Wet	3 (Seed)
		150-200 (fuelwood)
Marigold	Winter	35
Rajmash	Winter	8
Sunflower	Summer	5

the preceding option but the FSR team is controlled by extension department rather than be a part of the research apparatus and (e) a national FSR project controlled and funded from the centre but implemented by the state agricultural universities in their service areas.

The first option appears to be more appropriate in the present circumstances. A meaningful rapport between the client farmer and researcher with social assistance would help in introspection leading to a modification in research agenda and technology dissemination modalities concurrently. Adoption of this option will necessitate the lost structural adjustments in the present set up and at the same time would be less expensive one.

The successful incorporation of FSR principles and concepts depends on the will of the decision makers within the relevant institutions to take the matter seriously and to make the necessary structural, organizational, management and operational changes. Adjustement to teaching, changes in reward systems and reallocation of staff time is also required. A strengthening of the relationships between the departments within the University between the University and collaborating institutions, ICAR and the state agricultural system may also be necessary. A model for institutionalizating FSR within University structure is presented in Fig. 12.

5.6 Make-up and Functions of Component Parts

1. Farmer Groups : These groups should participate in all stages of research, evaluation and the extension process with the FSR core team and interact with team both as individuals and as members of target group.

2. Core FSR/E team. It should consist of small active team of scientists, mainly based in villages. A minimum of an agronomist, a social scientist and a livestock scientist are necessary. In addition, an extension liaison person is required and a senior scientist/ coordinator to act as a planner and link with the Farming System Research Advisory Panel. (FSRAP). All these team members should be drawn from existing Departments or Faculties of Agriculture, Social Science, Animal Production and Extension.

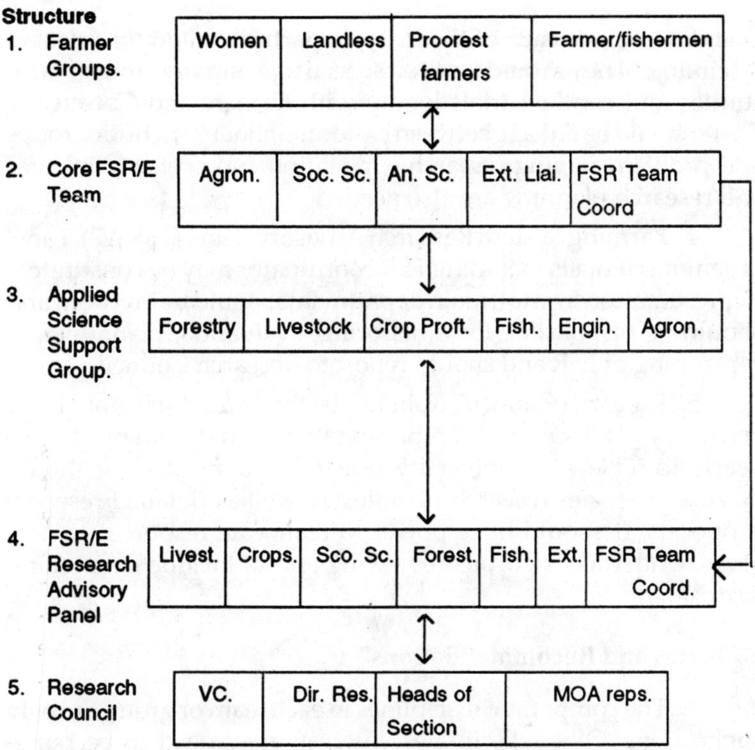

Fig. 12 : Model for Institutionalizing FSR within university structures

The primary functions of the team are : Preliminary and diagnostic surveys, agro-ecological and socio-economic analysis, priority setting, technology experimentation, development and evaluation. All to be carried out with farmer-colleagues. Additional inputs from Applied Science Support Group (ASSG) as needed. Basic research problems directed through ASSG. Feed back to on station research.

The output of this group is critical for the development of systems thinking and alternative modes of technology generation throughout the research system. Particular attention needs to be given to the process of team building, thinking and operation.

3. Applied Science Support Group : This group should include active, disciplinary scientists prepared to work in the field.

One drawn from each of the major departments and/or relevant disciplines. They should work on special topic surveys and no-farm studies and conduct trials jointly with farmers and Core team. There should be linkage between basic disciplinary scientific groups and problem solving research. On-station and on-farm linkages and research planning are also needed.

4. Farming System Research Advisory Panel (FSRAP). Panel of senior scientists including FSR coordinator may be constituted. Representation from all main departments should be involved and should be responsible for monitoring, evaluation, planning and replanning of FSR and should report to Research Council.

5. Research Council : It should be the normal statutory body consisting of all senior staff representatives and should meet twice yearly to review all University research programmes. It should discuss strategies research planning as well as detailed research proposals. It should have power to reallocate resources to new areas. Undertake rigorous reviewing and to include external researchers.

5.7 Notes and Recommendations

1. The component disciplines in each team or group are only suggestions. Other disciplines/interests may need to be represented.

2. It is considered not to be desirable to create a separate FSR Department.

3. The initial step is the creation of the small, core FSR team and the supporting groups. This team carries out initial diagnostic and research work. As priority areas are identified, additional teams or working groups can be established and disbanded as required. *e.g.* Soil fertility, fodder, ducks/small stock, land shaping technology, fishery systems. These groups could have the same relationship to the Advisory Panel as the FSR team.

There may be some difficulty finding suitable full time people for this team. It will be necessary in some places to redesign the responsibilities and reward systems of staff members. Perhaps all newly appointed agronomists, animal production scientists and social scientists should be required to spend a period of three years in FSR field work before their appointments are confirmed.

Permanent staff members should also be required to spend a minimum proportion of their time, *e.g.* a third, in village based research work.

4. Other teams involved in related ICAR projects, *e.g.* Lab to Land, Operational research, Dryland research *etc.* could also be fitted into this model as equivalent units to the FSR Core Team. This implies that many of these programmes, which are presently little more than demonstration/transfer of technology in orientation, could become more oriented towards adaptive research with full farmer participation. This would involve some changes, by ICAR, in the formulation, objectives and direction of such programmes.

5. It may also be highly advantageous to link the core team with appropriate NGO activities, and both utilise and complement the particular skills and experience of NGOs in community and rural development.

6. The discussion on the development of this model highlighted a key difficulty with the current Ford Foundation policy of giving modest support for the development of farmer participatory FSR at several institutions. Until and/or unless the Universities accept the principles and concepts of the approach and build FSR into their research environment and structures, the FSR programmes will continue to have low status, support and low priority allocation of human and other resources. Perhaps the combination of the ICAR changes in the allocation and use of their research funds and the continued level support from Ford Foundation on the institutionalization issues raised by this structural proposal, will result in positive changes that will help bring about an effective structure and process of farmer centred research.

SUMMARY

Technologies can be identified for sustainable development of high risk prone areas and that farmers have a critical role in modifying, disseminating and adopting these technologies. Similar arguments have been advanced in other recent publications (Chamber *et al* 1990, Conway and Barbier 1990). What is needed in the exploration of congruence among policy regimes, programmes and institutional collaboration which would encourage adoption of farmer first approaches on a large scale. This will imply that the FSR teams expand their horizon of operation by entering into collaborative action research programmes.

The foregoing case illustrations show the potential for synergy between development bureaucracy, grass roots organisations and the FSR team. Like all joint endeavour however effective tripartite collaboration calls for considerable investment of time and effort in clarifying roles of each partners as well as generating a realistic appreciation of factors which may act as deterrent to such collaboration. There will be synergy only when the agencies having differing comparative advantages collaborate to complement each other.

Recognising the advantages of FSR as a vehicle of agricultural development the modalities of its integration into the mainstream research and extension efforts need evaluation. As on the importance of farming system research approach should have the congnizence as a policy programme. What is required to enunciate the steps needed and the mechanism appropriate for integrating FSR into the national agricultural research and extension systems.

REFERENCES

Bottrall, A. 1990. "The Ford Foundation supported Farming Systems Research Programme in Eastern India". Paper presented at a Workshop on Linking Farmer Participatory Research to Policy Planning and Implementation, University of Kalyani, 6-9 March, 1990.

Bottrall, A. and John, K.C. 1990. "Adapting FSR to the Needs of Collective Decision-Making". Paper prepared for presentation in the Asian Farming Systems Research and Extension Symposium on "Sustainable Farming Systems-21st Century Asia : AIT Bangkok 1990.

Binswanger H. P. and McIntire, John. 1987. "Behavioural and Material Determinants of Production Relations in Land-Abundant Tropical Agriculture". Economic Development and Cultural Change 36:73-99.

Centre for Water Resources Studies 1990. "Assessment of Wasteland Problems in Humid Alluvial Regions—A case study in North Bihar", Centre for Water Resource Studies, Bihar College of Engineering, Patna.

OGIAR, 1978. Farming Systems Research at the International Agricultural Research Centres. Rome : FAO.

Conway, G.R. 1985. "Agro-ecosystem Analysis". *Agricultural Administration*. 20 : 31-55.

Conway, G. R. 1990. "Participatory Analysis for Sustainable Agricultural Development". Paper presented at the International Symposium on Natural Resource Management for Sustainable Agriculture, 6-10 February 1990. New Delhi.

Conway, G.R. and Barbier E. B. 1990. "After the Green Revolution : Sustainable Agriculture for Development, Earthscan Publication Ltd., London.

Gibbon 1990; I. P. Abroal and D.K. Singh. "Review of Ford Foundation supported Farming System Research/Extension Programme at Rajendra Agricultural University, Bihar, India". The Ford Foundation. Lodi Estate, New Delhi-110012.

Jha, K.C. 1990. "Jal Jamav Ki Sthithi Men Ikh Prabandh" (in Hindi) in State Level Sugarcane Development Workshop, 24-25 August, 1990, Patna.

Kumar, A; John, K.C., and Saran, S. 1990. "Sustainable Agricultural Development in Chaur : Linkages Among Policies, Transactions and Ecological Risk". Paper presented at a Workshop on Linking Farmer Participatory Research to Policy Planning and Implementation, University of Kalyani, 6-9 March, 1990.

Kumar, A; K. C. John, Ahmed Salman, Prahlad Mishra and S. Saran 1990. "Policy Implications of Farming Systems Research Experiences in Flood Prone Basins in Bihar". Best paper Awards 1990, Ist Asian Farming Systems Research and Extension Symposium, Bangkok, Thailand Nov., 10-12, 1990.

Salman, A; Kumar, A; Mishra, P. and Saran, S. 1988. "Caste and Class Specific Differential Role of Women in Farming Systems of Bihar with Special Reference to Vaishali Chaur". Paper presented at the Farming Systems Research/Extension Symposium, University of Arkansas, 10-12 October 1988, U.S.A.

Saran, S; Mishra, P; Kumar, A; and Salman, A. 1989. "The Role of Farming Systems Research in Sustaining Productivity and Profitability in Vaishali District of Bihar, India". Paper presented at Arkansas, 8-11 October 1989, U.S.A.

Saran, S; John, K.C; Kumar, A; Salman, A; Sinha, M.M. and Mishra, P. 1990. "Incremental Methods for Developing Integrated Farming System in High Ecological Risk Zones". Paper presented for Asian Farming Systems Research/Extension Symposium, AIT, Bangkok.

Chapter–6

Impact Assessment

6.1 Salient Features of Activities and Gains

On farm research over a decade in risk-prone environment and association with the farmers in implementing the research agenda in FSR project have given a new dimension to the approaches and thinking for inducing qualitative and quantitative changes in the farming system of the area. This has also led to the identification of priorities and dissemination of technologies with high promise in the background of socio-economic and farming environments of the recommendation domain. Farmers played significant role and the traditional knowledge and wisdom constituted the main input for modelling and developing research plans.

Considering the fact that agricultural out-puts are the backbone of the rural economy, the activities under the FSR project were mainly concerned with the improvement in production of various crops. Salient features of the activities pursued are presented here under to use for assessing the generation and acceptance of technologies out of several programmes implemented through farmers participatory approach.

— Environmental classification and characterisation is land system classification and identification of problems associated with different land types; physico-chemical properties of the soil; hydrological situations, land topography, mapping of *chaur* (basin land) system.

— Identification of viable cropping system, crop-varieties screening under different land situations and water regimes.

— Manipulation of risk-free period (post-monsoon) for assured crop production.
— Experimentation on nutrients management and intensification of crops in different land types.
— Identification of crops for mix-cropping in different chaur situations.
— Introduction of crops for increasing cropping intensity.
— Exploration of animal husbandry as a source of livelyhood; Rearing of milch-animal and goat; fodder development.
— Socio-economic characterisation, identification and categorisation of different groups of household economic status of different groups, caste categories, male and female farm labourer.
— The FSR team took-up issues highlighting inadequacy of interventionist policies and practices for water-resource management, processing of agricultural produce.
— Organising workshop, seminars, Kisan Gosthi, presentation of problems and solution by the farmers and scientists.
— Activating various Government, N. G. Os and other organisation for participation in assessing the problems and developing viable solution.
— Promoting a series of Farmer Participatory Policy Dialogue Workshops, bringing together farmers, policy makers, administrators, implementors, N. G. OS and agricultural researchers on a platform. In addition the formation of Pani Panchayats or water associations aims at laying the organisational foundation for co-ordinated water management across *chaurs*.

6. 2 Lessons Learnt

The rice dominated cropping systems (deepwater rice, DWR) and mixed cropping of rice in the high-risk zone could be diversified incrementally. Our experience of working with farmers suggests evolving an ecologically differential land-use plan (Table-19).

Instead of circumscribed by the entry point strategy it is critical to be flexible and monitor the coping strategies of the local community to devise an integrated farming system (IFS). Even successful FSR programmes have rarely been able to take into account the coping mechanism of farmers.

— Sensitivity to the farmers' informal experimentation (Biggs 1980; Richards 1985), evaluation (Ashby and Pachico 1989) and indigenous resource classification system (John 1989) enables FSR practitioners to draw a general outline of an IFS around existing micro-ecological niches. Such a broadened system technology could then be tested or developed by working with larger group of farmers and subsequently refined.

— The key research objective during the initial phase should be to develop a better appreciation of the current thrust of farmers experimentation to create newer options for managing their ecologically constrained resources, and then creatively exploring with farmers better ways of exploiting these options to meet the multiple objectives of the community.

— As a corolary, it is important to know farmers' preferences and judgements about the conditions in which basic components of the technology can not only work but also be well integrated into existing systems. For instance, with *Sesbania rostrata* and summer rice varieties, it was critical initially to understand the conditions under which they can be successfully grown. A configuration of factors, such as residual moisture, onset of monsoon and flood, water-receding pattern, etc., would govern the integration of newer component into the system (Table 19).

— Obviously, instead of identifying optimal conditions for an IFS-there could be no optimal condition or recommended set of practices for a complex, dynamic and risk-prone environment as that prevalent in the *chaur*-the objective should be to identify the range of flexibility in management of a particular component of IFS. For instance, through the FSR approach, it would be possible to define the range of water level and the speed of its receding in choosing between puddle wheat or late sown wheat.

— It is also important to be sensitive to the ways adopted by farmers to make the component technology work within

Table 19 : Integrated Farming Systems for micro-environment of *Chaur.*

	Niche	*Kharif*	*Rabi*	*Summer*
A.	Bheeth	Maize/rice	Fingerling rearing Tobacco and vegetable or spices Maize rajma Oilseed *Marigold*	Maize Mungbean Vegetable Sunflower
B.	Uncha Chauri I	Rice (transplanted) Sugarcane	Wheat Bakla	Rice (transplanted) Maize + deep water rice Sugarcane + Mungbean
C.	Neecha chauri I	Rice (transplanted)	Wheat Barley	Maize + deep Rice (Transplanted)
D.	Neecha chauri II	Rice (transplanted)		Rice (transplanted) Maize + deep water rice
E.	Neecha chauri III	Rice (transplanted) Deepwater rice * jute Deepwater rice + S. rostrata Mungbean + S. rostrata Jowar + S. rostrata	Integrated with aquaculture	Rice (direct seeded) Maize + deep water rice
F.	Sarkurba	Rice (transplanted) Deepwater rice + jute Deepwater rice S. rostrata + Jowar + S. rostrata		Rice (direct seeded) Maize + deep water Maize + deepwater rice
G.	Leebh	Deepwater rice + jute Mungbean + S. rostrata Deepwater rice + S. rostrata Jowar + S. rostrata		
H.	Macha- garah	Deepwater rice + jute Deepwater rice + S. rostrata		

* Integrated with bee keeping.

the existing production environment. Tolerance for deviation and divergence is the key to success. Summer (transplanted) rice could not have fitted into the chaur system had the researchers not respected the individual farmers' right to decide the date of transplanting.

— Another important issue that is still being explored is the conditions that make the systems technology attractive to different groups of farmers : resource-poor tenant farmers vis-a-vis peasant cultivators, the fishing community vis-a-vis peasant cultivators.

— Developing an IFS for complex, diverse and risk-prone environments is a long-term process. We have just been able to draw an outline of such an IFS for the flood-prone basins (*chaur*).

— It is also crucial to keep in mind the resistance from on-station researchers. FSR practitioners in the present case continued exploring various plausible options suggested by farmers in their own fields. Emperical evidence was gathered and rigorously used to persuade on-station researcher colleagues to be sympathetic to on-farm research and to the farmers' reactions to it.

However, unlike in conventional on-farm trials and mainstream research, the FSR team was hesitant to recommend a best-bet strategy for the interlinked *chaurs* of northern Bihar. Nor they attempted validation of an IFS for all *chaurs*. They suggested that lessons learned from the FSR experience could enable agricultural researchers to draw a general outline of an IFS appropriate to ecologically risky regions. Through this approach it is also possible to generate a range of options that farmers can try to adapt in their own micro-environments.

FSR practitioners do not suggest that they have found the ultimate solution to developing an IFS, but they concur with other FSR practitioners that the outcome of an FSR programme in the short term is best measured by farmer-to-farmer diffusion and continued interest among farmers in experimenting with components for integration of FSR in their own systems.

6.3 Diffusion of New Technological Options

On-farm research over a number of years resulted in identification and dissemination of a number of promising technological improvements.

(a) *Rabi* (winter season) wheat variety HP 1490 was adopted by the farmers of the area. Within 3 years of its introduction the variety covered almost 80 per cent of the total wheat area. Subsequently it spread to the nearby villages. It was heartening to observe that the diffusion of the technology including seed production was entirely carried out by the farmers.

(b) Seeds of *Sesbania rostrata* were received from International Rice Research Institute, Manila and were tried in the paddy fields. *Sesbania* is a legume which has nitrogen fixing nodules not only in its roots but also in stems and other arial parts and is thus an efficient fixer of atmospheric nitrogen.

- Since the farmers in the *Chaur* lands use commercial fertilizers, the FSR team proposed to introduce *Sesbania* as a green manure crop to be ploughed back into the soil after the floods recede and the land is ready for transplanting paddy.
- The farmers were not initially enthusiastic. Its cultivation would involve extra work and they were not convinced of the benefits. However a few farmers reluctantly agreed to try it on the boundaries of a small part of their land in an area which remains submerged for a long time after flooding and hence no other crops can grow. There in lies a success story.
- The *Sesbania* plants grew vigorously, put-forth plenty of foliage and attained heights of about 10 to 15 feet in a relatively short period of 4 to 5 months, even under medium and deep water conditions. The farmers noted that the boundaries of their submerged plots were demarcated by the waving foliage.
- Pretty soon they found other uses of the *Sesbania* plants. For example, their dried sticks proved good as thatching rod for the roofs and for the construction of

grain storage bins. The stouter sticks were easy to place for marking plot boundaries and stoutest for building poles. The green part makes excellent fodder for the goats and because the goats can be tethered, reduces the amount of damage free-ranging goats do. The sticks could be tied to form a reef barrier for catching fish or inter-woven as a barricade. They also greatly augment domestic fuel supplies. The Entomologists also found the dried stems more resistant to weevil attack than the conventional bamboo reeds.

☐ The farmers continued to experiment. For example they tested with researchers support, Sorghum/*Sesbania* mixers as a fodder crop and rice/*sesbania* mixers for nutrient management and to stagger the cultivation season. They also experimented with researchers help rice-fish farming in barricated tanks. In these trials researchers first recommended constructing the barricades once the chaur was flooded but farmers observed that then predator fish had a chance to enter the tank and eat the fingerlings. Together they are now working on barricading whilst the land is still dry.

☐ The seeds of *Sesbania rostrata* are dormant for a year. However another species *Sesbania acculata* does not have this problems and was also introduced by the researchers. The farmers mean-while learned to (a) break dormancy by wrapping the seeds in the leaves of another plant and soaking (b) preserve the seeds of rostrata for use in two year cycle (c) use aculata seeds in the intervals as supplementary supply. Thus they have developed a robust mechanism for continuing supply of seeds for their own use and for sale. Cultivation of the *plants* has spread widely.

☐ Farmers feel that *Sesbania* plant is a boon to them. Their women do not now have to walk miles to fetch firewood. They can utilise extra time for many other household activities. They did not accept *Sesbania* as a green manure plant but the many uses of the plant attracted them.

(c) Summer rice cultivation attracted the farmers. They found it to be a reduced risk strategy for shallow and intermediate water situations. But provision of irrigation and its higher cost of cultivation did not encourage them to adopt this technology in a big way. They made innovations and delayed the seeding time. This strategy helped them in reducing the number of irrigations to the summer rice crop thereby minimising the cost of cultivation. Some of the enterprising farmers introduced the summer rice varieties in chauri lands (inter-mediate lands) during kharif season (Wet season) and harvested successful rice crop. IET 3279 and Pusa 2-21 varieties were selected by them from the Summer rice variety trials. These varieties are now popular among them as summer as well as kharif crop.

(d) A large number of other technical options were also researched. These have encouraged greater diversity and also intensified the farming system. These are as follows :

(i) Mixed crop of rice + jute was found to be profitable and sustainable for intermediate deep and deep water lands

(ii) Inter-cropping with tobacco

(iii) Puddled wheat cultivation

(iv) Introduction of new oilseed e.g. Sunflower and pulses varieties etc.

(v) Improved agronomic practices in various crops

(vi) Improved Rice-fish culture

(vii) Marketing of crabs and snails

(viii) Fresh Water prawns

(ix) Floriculture specially marigold for sale in Patna

(x) Introduction of fodder crops e.g. S. S. G

(xi) Ducks in the Chaur

(xii) Improved goat breeds

(xiii) Bee keeping

(xiv) Sugarcane cultivation

6.4 Sustainability, Stability and Productivity

Diversified portfolio of enterprises ensure sustainable development of high risk eco-systems by increasing the diversity. Farmers of region have evolved a mixed cropping system such as rice-mungbean-sorghum and different sowing and crop establishment methods for hedging against risk. In its attempt to develop an integrated farming system, the FSR team had been exploring integration of crops like sugercane, vegetables, flower and non-conventional crops in the chaur mentioned earlier. The ultimate objective was to develop a sustainable farming system suitable for the local condition which is stable and profitable.

6.5 Achievements

(1) The establishment of a good working relationship between Researchers and different farmers groups.

(2) The development of a better understanding of Farming System Research and farmer participation by the Scientific community and the realization that such approaches are necessary because of limited application of conventional research methods.

(3) The appreciation of the need for a team approach among the scientists.

(4) The experimentation with farmers gave precision to the knowledge of the scientists, leading to the better understanding of the micro-environments and problems of the farming system of the area. It also lead to the development and dissemination of number of viable options.

   (a) Diffusion of wheat variety HP 1490.

   (b) Cultivation of *Sesbania rostrata* in shallow intermediate and deepwater situations. The introduction of this crop has opened up a new series of options and has stimulated innovations by the farmer *e.g.* fuel wood, boundary markers, mixed crop etc.

   (c) Summer rice has been found to be a reduced risk strategy for shallow and intermediate deep water situations.

   (d) Mixed cropping of rice + jute was found to be both profitable and sustainable for inter-mediate deep water situations (70% land area).

(5) A large number of other technical options were researched

which caught the immagination of the farmers. These efforts encouraged greater diversity, intensity in the farming system. They include intercropping different crops with tobacco, rice varieties, puddled wheat, oilseed variety pulses variety, vegetables improved agronomic practices, rice fish culture, fresh water prawns utilisation of crabs and snails, floriculture especially marigold, improved goat breeds, fodder crops, ducks in *chaur*, bee-keeping etc.

(6) Multidisciplinary as well as multi-institutional approach in the development of FSR programme.

(7) Last but not the least, development of the irrigation facilities (for the first time in the area) with the help of lift pumps by the farmers themselves. When the FSR project was initiated in 1984 the irrigation facilities were negligible. In 1994 more than 10 lift pumps were operational. The farmers are very much aware to derive the maximum benefit but the new technological options made available recently specially for dry season (summer and *rabi*). There is plenty of water available in the *chaur*. What is needed is to develop the irrigation facilities.

6.6 Strengths and Weaknesses of the FSR Programme

A Strength

The most notable strength of the programme was the close rapport between the FSR team and the farmer groups of the area. This had generated confidence and mutual motivation to develop farming systems and livelihood.

Success in generating alternative technologies was achieved by a willingness to build on farmers knowledge, experimentation and participation in research extension and evaluation.

An efficient system of generating new ideas and testing them quickly among collaborating farmers has come into operation. The FSR team has learned rapidly from early approaches and brought the farmers more into research decision making.

The social science group carried out some valuable gender-sensitive research on the chaur communities and have developed a good inter-action with women's group in the area.

B. Weakness

(1) The Concept and approaches of FSR were not fully understood by all collaborating scientists.

(2) The essential linkages between commodity and disciplinary research and farming systems research and between on-station and on-farm research were not present. The university's research programme is currently structured in a centralized, compartmentalized manner that does not encourage effective feed back from FSR work to the disciplinary, basic research programme.

(3) There were limitation in the organisational structures and transport constraints.

(4) Extension linkages were weak.

(5) Methods of inter-disciplinary research needed improvement.

(6) There was some constraint in the access of funds.

(7) Well defined researchable issue for poorer farmers were not apparent.

(8) The whole process of review, analysis, implications and replanning was not very systematic.

6.7 Recommended Lines of Action for the Improvement in the FSR Programme

(1) More effort is needed to develop the understanding of the importance of FSR concepts, approaches and methods within the collaborating departments and institutes. The links with different types of research also need attention.

(2) The implications of starting with the farmers and developing a research programme within the socio-economic environment of the farming family need to be discussed at a regular seminar series of all relevant staff.

(3) More active thought needs to be given to the institutionalization of FSR. This may require a change in structure and organization and a series of actions from the researchers.

(4) The FSR team should have wide responsibilities for all interventions and research on-farm and operate in a gener-

ally inter-disciplinary manner. Farmers must be fully represented in the planning of group activities.

(5) The significance of the social science surveys should be discussed. Groups of rural people should be identified and implications of research for the very poor groups should be realised.

(6) Better linkages between on-station and on-farm research are necessary. All commodity or disciplinary scientists who produce new technologies for farmers should be required to test such technologies on realistic on-farm environments before making recommendations for extension.

(7) Better linkage between the FSR programme and the ICAR programme and extension is necessary. Greater co-ordination of Govt./University and NGO efforts is needed. The FSR group need to have more interaction with developmental agents. This will allow the follow up of the initial findings of the research and allow the team to focus on new priority areas of research.

(8) Monitoring, evaluation and impact analysis needs to be carried out by an inter-disciplinary team.

(9) Emphasis is needed on indigenous technological knowledge, experimental methods, renewable resource based methods. Greater emphasis is needed on the sustainability of the systems issue.

(10) Some examination of loan/credit/revolving fund mechanism available for the very poor is necessary.

(11) At the university level the FSR programme should be discussed thoroughly and the FSR team should be encouraged to participate actively in the review of disciplinary research work on the issue of relevance of research programmes intended to help poorer farmers.

(12) The concepts and approaches of FSR need to be incorporated into teaching programmes.

(13) The characterisation of the wider agro-ecological and hydrological situation is necessary. A study of changes in

hydrology of *chaur* in recent years is needed and the implications for future land and water management in the region.

(14) Further research on pond digging technologies in the chaur for fish and tree planting on the embankments is necessary.

(15) Assessment of FSR work on regular basis is necessary and reward need to be given to good researchers on the basis of their performance.

(16) Much more work on the presentation of findings is required. Not only the technologies that are produced but the process of research and extension needs to be presented. This information should be invaluable to other teams who are at an early stage of development.

(17) A longer term study of individual households of systems representing the main socio-economic groups of villages.

(18) Emphasis be laid on the resource based research than the commodity or discipline based research.

SUMMARY

The most notable impact of the programme was the close rapport between the FSR team and the farmers groups. Success in generating technologies was achieved by a willingness to build on farmers knowledge of experimentation and participation in research, extension and evaluation. An efficient system of generating new ideas and testing them quickly among collaborating farmers has come into operation.

The social science group has carried out some valuable gender-sensitive research on the *chaur* communities. In addition they identified and classified different socio-economic group, classes, households etc.

Environmental classification and characterisation, land system classification and identification of farmers problems,were useful gains. Promotion of Farmer Participatory Policy Dialogue Workshop leading to the formation of *"Pani Panchayats"* for coordination water-management activities of the inter-linked *chaurs* by the farmers was an important achievement. No where in India Policy dialogue exercise was held before.

The usefulness of FSR for agricultural development was recognised. It was felt that there is need and scope for integration of FSR into mainstream research and extension activities.

REFERENCES

Gibbon, D; I. P. Abrol and D. K. Singh (1990). "A review of the Foundation supported Farming System Research Programme at Rajendra Agricultural University, Bihar". The Ford Foundation 55, Lodi Estate, New Delhi 110012, India.

Raman, K. V; J. Jiggins (1994). "On Farm Research in Eastern India. A Deccenial Review". 55 Lodi Estate, New Delhi-110 012, India.

Chapter –7

Future Prospects

7.1 Important Issues

Three important issues merit consideration are (1) Sustainability of FSR (2) Sustainability of natural resources (3) Project orientation to FSR. Out of these issues emerge the following points which are important for future of the FSR.

— The FSR concept can not be institutionalized unless recurrent costs can be met. This is empeded by government research and extension budgets that leave few resources for carrying out on-farm activities. External support for FSR must provide incentives for public and private funding of research and extension and must ensure that the host institution research organization develop a capability to assume FSR recurrent costs.

— If properly implemented the FSR can offer an excellent vehicle for addressing the sustainability of natural resource base. The challenge is to ensure that sustainability initiatives involving FSR core characters are not undermind.

— Implementation of FSR has not been facilitated by an assistance made (project) that provides assistance to 3-5 years span; indeed the limited impact of FSR project to a certain extent is due to its short life-span. Success in research and institutional development requires a longer time-frame and this is no less true for FSR.

— FSR is not a substitute for conventional research but can be instrumental in accelerating the speed with which the technologies are developed and transferred. But this process is not aided by a short-term orientation to agricultural research in general and FSR in particular. Support needs to be sustained over long-term (15 to 25 years). The challenge for future is to address the constraint to FSR more effectively. This may be done by ensuring :

(a) The FSR core characteristics are systemetically built into the technology development and transfer methodologies.

(b) The agricultural research/extension project should provide a means to remove or relax the operational constraint that can impede the implementation of FSR.

(c) The project assistance to relax core and operational constraint to FSR as not undermind by generic constraint. The FSR project has tended to focus on the food crops paying little assistance to other commodities. There is scope for diversification and place emphasis on cash crops, animal husbandry, fishery, horticulture *etc.*

— Research mandates have caused FSR practitioners to focus improving production technology. (Primarily food crops) as the end rather than a means. The approach should be to include increased farm family income also into the design of FSR and focus on the farm and farm family as a system.

— FSR practitioners should address institutional constraints to adoption of the technologies being developed. Farmers often cannot adopt such technologies unless they also have access to such agricultural support services as credit production inputs and markets. Here social scientists of the FSR group have to place greater attention on identifying means to remove or relax institutional constraints that impede farmers access to agricultural support services.

— FSR teams can play an important role in linking research and extension by working with farmers and extension to test-adopt technologies derived from research and with researchers to provide feedback to establish research priorities.

— FSR impact on technology development and transfer will be negligible until research and extension personal work out a joint strategy to institutionalize FSR methodology in research and extension programmes.

One of the problems encountered in implementing the FSR project is due to limited knowledge and understanding of the requirements for implementing this concept. FSR when properly implemented can strengthen the technology development and transfer capability of agricultural research and extension systems.

7.2 Challenges Ahead

The challenge is to integrate FSR into technology development and transfer methods. FSR explicitely recognizes the need for links among farmers, extension workers and researchers and defines essential conditions (FSR core characteristics) for increasing the impact of donor, government and private investment in agricultural research and extension.

However such impact can not be fully realized unless development assistance also addresses the various operational constraints that can impede institutionalisation of FSR. This will require a long term commitment to institutionalize technology development and transfer systems responsive to the problems faced by small holder farmers in the developing countries.

7.3 Future Directions for FSR

The guiding items for future directions of FSR are :

Broadening scope : FSR will continue to offer a useful paradign for research on technology generation. The scope of FSR may be broadened to integrate more farm enterprises, and to work more readily at the family, village, agroecosystem or watershed level. Under some conditions, FSR practitioners may need to pay more attention to nontechnical constraints or agricultural policy.

Affecting policy : FSR practitioners should pay more attention to policy issues, and should target policy makers as clients for research results.

Promoting equity : FSR procedures should be among those best suited for generating technologies appropriate for resource-poor farmers in complex, diverse environments. However, research focusing on resource conservation can generate recommendations that endanger the economic interests of low-income farmers. Researchers must be careful to weigh the likely distribution of benefits arising from resource-conserving innovations.

Incorporating external factors : In view of increasing effectiveness of FSR the practitioners have to give adequate attention to transaction costs, common property resources, pricing policies, and other external factors that can cause private and social costs or benefits to diverge. Policies to address land degradation, agroecosystem diversity, and links between agriculture and global pollution will have increasing impact on farming systems. FSR practitioners should be prepared to deal with such external factors.

Setting priorities : It may prove impossible for FSR programmes to expand in all directions at a time. Just as researchers must set priorities among technical research themes, so must they set priorities among directions for expansion. This will require careful matching between specific FSR approaches and institutional restrictions, environmental conditions, etc.

Involving farmers : The role of farmers in FSR should be expanded. Sustainable practices often have their roots in farmers' experiments and the recommendations of innovative farmers. Farmers can also be used to train other farmers and to collect and report data.

Working for sustainability : Researchers and research management must understand and examine agroecosystem diversity, links between agricultural enterprises, and the vulnerability of agricultural systems to climate changes. FSR offers methods for assessing changes in farmers' circumstances over time, and should be especially useful in recommending strategies for sustainability.

Integrating systems : As sustainability becomes a higher priority for FSR, the notion of "biological restructuring" of farm enterprises increases importance. Restructuring emphasizes nutrient and energy cycling, and requires integration of new enterprises. An integrated system maximizes the recycling of waste and by-products between different enterprises within the farm.

Building models : Conceptual and mathematical models must be developed and improved. Formal modeling can be complex and expensive, however, researchers should take care not to become overly reliant on formal models, and modelers should be aware of the need for substantial farmer input in model construction.

Assessing impact : There has been relatively little documented impact of FSR at the farm level. Donor fatigue is already apparent. Although methods of monitoring and evaluation are well developed, few FSR practitioners are familiar with them. Far more, attention must be paid to assessing the impact of FSR activities.

Emphasizing gender : FSR practitioners should understand and emphasize the role of women, especially in ecological restoration and in income generation when food or cash is scarce. This improved understanding must be brought to bear on research design and evaluation. Research managers must target women as beneficiaries of FSR activities, and must work to increase female participation as FSR professionals.

Strengthening research links : Links between FSR researchers and disciplinary and commodity scientists must be strengthened. All scientists working on agricultural technology generation should be encouraged to work from a systems perspective. The challenge is to convince disciplinary scientists of the value of FSR tools and concepts.

Working with extension : Farming systems research must link more directly with extension services. This will require the development of simplified, streamlined procedures—a minimalist approach. Farmers themselves should be called on to perform more extension functions through farmer-to-farmer training activities.

Teaming up with NGOs : When research resources are limited, nongovernment organizations (NGOs) can be of immense help in directing technology generation activities towards small farmers. Moreover, NGO participation can foster creativity in the development of FSR procedures. Governmental FSR programmes should strengthen their links with NGOs.

Improving training : The new directions suggested for FSR will require FSR professional to develop new skills. Universities and agricultural education systems must be prepared to meet this challenge.

SUMMARY

The great number of FSR approaches, concepts, procedures and terms developed and used all over the world is evidence of the richness of more than two decades of experience in FSR. Some practitioners feel that the actual impact of FSR has fallen short of expectations. Consequently more attention should be paid to the efficiency of FSR and especially to streamlining tools to hasten and increase FSR impact (Future direction of FSR in Asia 1990).

Only modest efforts have been made to assess the effect of FSR activities on farmers, consumers research institutions, and the environment. These efforts should be strengthened. Farmers have been and should continue to be the primary users of information generated by FSR. However, the informations should be packaged for use by policy makers, researchers, administrators and other.

REFERENCES

"A review of AID Experience : FSR Project 1975-1987". AID evaluation highlights No. 4 Centre for development and evaluation, U.S. Agency for International Development (AID). Washington D. C. 20523.

Future directions for FSR in Asia (1990). International Organizing Committee; Asian Farming Systems Symposium 1990, Island Publishing House Inc Sta Mesa P. O. Box 406, Manila, Philippines.

Raman, K. V and Balaguru, T. (1971). Farming Systems Research in India : Strategies for implementation Proceedings of the National Workshop Nov., 25-28, 1991 pp NAARM, Hyderabad, India.

Raman, K. V and Jiggins, J. (1994). "On-farm Research in Eastern India". A Decennial Review 1984-1994. The Ford Foundation, 55, Lodi Estate, New Delhi, India.

Index